Turoe and Athenry Ancient Capitals of Celtic Ireland

Fr. Tom O'Connor

Edited by Kieran Jordan

Published by Tom O'Connor.

Available From:

Kieran Jordan
Strawhall,
Fermoy,
Co. Cork.

Tel: 025/31559

ISBN 0-9544875-0-8

Front cover shows: The Turoe Stone

Front Cover design: Kieran Jordan
Photography: Tom O'Connor and Kieran Jordan
Maps drawn by Tom O'Connor and Kieran Jordan on an O.S. base map.

First published 2003 by Tom O'Connor.

Printed by Litho Press, Midleton.

This book is dedicated to the people who contributed their memories and knowledge to its content, and to the people who, over the centuries, preserved oral history so that one day it would gain recognition.

Acknowledgements

I express my gratitude to the many individuals, authors and institutions without whose assistance, advice, knowledge, inspiration and encouragement this work could not have been accomplished. First of all I want to thank countless land owners throughout the areas covered by this book. I must acknowledge the massive role played by them in directing my steps to follow up the routes of the various sets of linear embankments or ancient roads running through their, or their neighbour's, lands. They kindly provided an enormous fund of useful information, folklore and ancient traditions of one kind or another. Without their assistance, the recording of this material would never have been possible and might have been lost forever. Only a few are mentioned in this short book, others are mentioned in the book Hand of History – Burden of Pseudohistory, which will follow. Their store of local knowledge and folklore made the gathering of this material a fascinating adventure. Many have died without seeing the results of their collective memories in print. I am deeply indebted to all of them.

Many distinguished academics were generous with their advice and encouragement; Professors and authors Francis John Byrne, Daibhi O Croinin, Barry Raftery and Etienne Rynne kindly read my manuscript. Their wise and valuable council shaped, toned, shortened and simplified the text. I thank them for their enthusiastic encouragement to press on with its publication. Indeed Francis John Byrne's 'Kings and High Kings' and the late T. F. O Rahilly's 'Early Irish History and Mythology' were a constant check on the early history and Dindsenchas literature projected in this work. In researching the material for this work and reaching the conclusions arrived at, I have drawn on the research work of these and other academics scattered in many articles, journals and monographs. Although not recorded here, these sources are acknowledged in the text, footnotes and bibliography of the book Hand of History – Burden of Pseudohistory.

I own my thanks to the National Library in Dublin, the University College Library in Galway and the Map library at Trinity College Dublin for permission to consult hard to find books, journals and articles and browse over old maps. To all of these I am enormously indebted. I would like to thank Michael Lawless for help during the work, Sean Hanley for proof-reading the book and for helpful suggestions and Tom Varley for constant encouragement. Thanks are also due to William Finnerty who has built a web site for the book at www.turoeandathenry.com.

Finally, I would like to single out in particular my editor, Kieran Jordan, a Scientist with Teagasc, whose keen editorial eye shaped this shorter version of my main work pending publication, namely 'Hand of History, Burden of Pseudo-History'. His infectious enthusiasm for the project came just at the right moment when my own was beginning to flag. He kindly undertook the laborious task of proofreading and correcting the draft of my main work. He saw the immediate need for a shorter popular version. I am particularly indebted to him for generously taking upon himself the tiring task of going over the main work once more to extract and edit the material for this short popular version. It is due to his efforts that this book is now in print.

Special thanks to the following for permission to reproduce maps:

Ordnance Survey Ireland
Trinity College Map Library
National Mounements and Architectural Protection Division

Turoe and Athenry; Ancient Capitals of Celtic Ireland.

Preface

The primordial influence for this work goes back to my boyhood days around the fireside when neighbours talked the long nights away. They often discussed the purpose of the "double ditches" so widespread across the Galway landscape and the reason why a local "ancient city site" in the Turoe/Knocknadala complex were never mentioned in the history books. One loquacious visitor, the late Martin Finnerty, local history addict and storehouse of traditional lore, author of 'Punan Arsa' and rate collector, enthused about the "Fir Belg dykes" and the "outer dun" of Athenry, a 'muc cladh" as he called it, running for miles across the countryside. He spoke of Athenry as one "Celtic Capital" surrounded by these embankments and Turoe as another. These fireside chats fired my young imagination and years later drove me to satisfy my curiosity. As I went on to explore these linear embankments and seek their purpose I was fortunate to meet Patrick Keane of Fearta on Turoe's north slope. As he walked me across the length and breadth of Turoe hill he pointed out vestiges and surface traces of the embankments surrounding the entire hill, and recounted the traditions about these passed on by the old people of his young days. He referred to other embankments further afield and named elderly persons who could help me locate these. With this information and that of the "ancient city site" which I explored, I knew I was on to something very important. What exactly I did not know. It led to exciting research work. From that day, some 40 years ago, it was like a ripple effect as I moved further and further afield across the entire Galway landscape from one elder to another in chase of what eventually I came to know as the shadow of the mythological magic boar. The purpose of the linear embankments and the origin of their Fir Belg builders became almost an obsession. What began as a pastime soon became an all-consuming quest touching the very heartbeat of Iron Age Ireland, 2000 years ago.

Oral tradition was of immense importance to the Celtic people in prehistoric times. When history was first recorded in written documents about the 7th Century, the work was carried out by scribes of the larger ruling families. What was recorded about the Ireland of 2000 years ago was totally at odds with the oral tradition. But over the years, since we have a tendency to believe the written word, this is what became accepted as history. This written version of history in archaic documents like the Book of Leinster and the Book of Armagh was thought to be correct at the expense of oral tradition and legendary history in the Dindsenchas documents. And so, pseudo-history was born. To account for the differences between the written and oral history a certain mythology was created around Ireland's Iron Age history of 2000 years ago, to the extent that it is now difficult to distinguish between fact and fantasy. However, oral history from this time has lived on and has been passed through families for generations around the fire. In addition, archaeological evidence has also survived the passage of time. Both the oral tradition and archaeological evidence in Galway are totally different than the Irish Iron Age history found in books. Because of changing lifestyle and changing agricultural practices in the 20th Century, both the oral traditions from 2000 years ago and archaeological evidence of Galway are in danger of being lost. The importance of retaining and revitalising these is essential so that a proper understanding of what really happened in Ireland 2000 years ago can be achieved.

This book revitalises the oral traditions, reconstructs archaeological evidence and delves into archaic documents for scraps of unaltered written evidence for the existence of Celtic Centres of Power in Galway. The conclusions are quite clear and shatter the accepted 'truth' of Iron Age Irish history suggesting a total re-evaluation of the origins and history of Celtic Ireland.

It is not the intention of the book to detract from the present understanding of Irish history, although by its nature this is inevitable. It is intended to give recognition to the people and places that were important in Iron Age Ireland and to the people who carried the oral traditions over the centuries so that they could be resurrected almost 2000 years later.

Table of Contents

Glossary of Terms

Oppidum	An enclosed Population Centre of a Celtic Tribe
Belgae	Also called Fir Bolg or Fir Belg they were a group of north European Celtic tribes
Ramparts	Another name for the linear embankments
Iron Age	From about 500 BC to 500 AD. The book deals with the second half of the Iron Age
Fosse	The drain between two banks
Inner Ward	The area enclosed by the inner-most set of embankments
Outer Ward	The area enclosed by the outer embankments
Ptolemy	A 2nd century Greek geographer, who left a record of Irish political geography at that time
La Tene	A particular period during the Iron Age for which a type of Celtic Art is well-known
Dindsenchas	Legendary history of famous places, written in about the 11th century
Cruthin	The native people of Ireland at the arrival of the Belgae in the 1st century. They were driven north to Ulster
Pseudo-history	False history based on fictitious and invented documents

Overking	Each small area had its own King. The Overking had control over the local Kings
Vestiges	Remaining parts or fragments the linear embankments
Souterrain	An underground cavern, built for protection or storage of food
Other World Abode	The Celtic people were not Christians and worshipped many gods. The people believed that these gods lived among them at an underground Other World Abode.

Timeline

YEAR	EVENT		AGE
			Medieval
		Fall and suppression of Turoe	
500			
			Christian
	Saint Patrick	*Niall of the 9 Hostages*	
400			
			Pre-Christian
300			
		Cormac Mac Art	
	Art		
		Conn of the 100 battles	**Iron Age**
200			
		Ptolemys record of Ireland	
	Queen Medb		
100		*Dela arrives in Ireland, establishes Turoe as his Centre of Power*	
0		*Birth of Christ*	
- 50	*Roman invasion of Britain*		
			Bronze Age

Introduction

Extensive research, carried out over a forty year period and based on ancient manuscripts, old Ordnance Survey (O.S.) maps, archaeological evidence, townland names, legendary history and other folklore stories, has led to the discovery of a vast Belgic (Fir Belg) Oppidum (Population Centre) which goes back to the origins of Celtic Ireland and which, although incredible, is certainly worth further investigation by archaeologists and historians.

It is generally accepted that Tara, Co. Meath, was the seat of the High Kings of Ireland from the earliest times. This assumption is based on the writings of 7th Century scribes who were funded by the Ui Niell Dynasty. Other medieval writings, like the Book of Leinster and The Four Masters have more or less followed the earlier writers. Irish historians of the 20th Century have shown that the 7th Century writers were economical with the truth, in order to glorify Tara and its Kings who supported them financially. Reference has frequently been made to the fact that early Irish history needed a total reconstruction. E. MacNeill stated in 1904, that *'the received framework of early Irish history was an invention of chroniclers from the 9th century on, working assiduously for the glorification of their patrons.'* By the 1960's the work of a whole generation of scholars had exploded the basis for popular assumptions about early Irish society. The idea that false images of Ireland's past were undermining its present and mortgaging its future was quite current by 1977. Roy Foster noted *'that the accumulated findings of historians from the early 20th century show the High Kingship of Ireland at Tara of Meath did not exist before the 9th century A.D'*. William Moody and Dudley Edwards embarked on a scheme to bring about a *'revolution in the aims, method and style of Irish historical writing,'* concerned for *'a purer, more scholarly kind of Irish history'*. The course of Irish history was to be systematically reconstructed by a new generation of scholars highly trained in the methods of the professional academic historian. A vast amount of

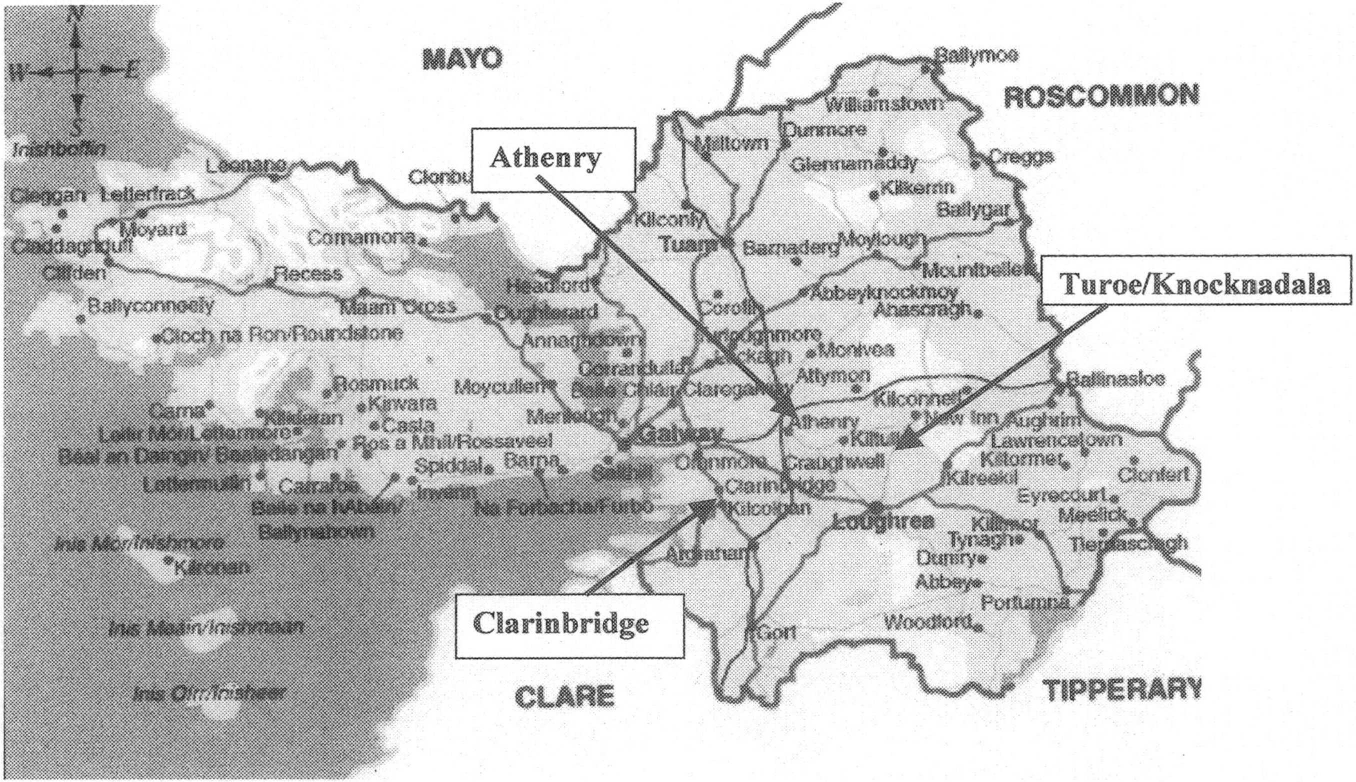

Location map showing the relative location of Turoe/Knocknadala, Athenry and Clarinbridge.

superb academic work has resulted from this plan. Yet, major areas remain untouched or have so far been overlooked. Turoe/ Knocknadala/Athenry is one of these areas.

This book examines the accepted version of Irish history and compares it to oral versions in Galway. With the aid of the archaeological remains, oral history and the findings of modern historians it concludes that the glorification of Tara occurred at the expense of the Chief Capital of Celtic Ireland (outside Ulster) which was situated at Turoe/Knocknadala/Athenry in Co. Galway. By reference to modern historians and ancient manuscripts this book shows the enormous importance of this newly discovered Turoe/Knocknadala/Athenry Belgic Oppidum to the early history of Connaught, addresses this area and examines closely the ancient records, oral history, archaeology, recent historical findings and other evidence from that area.

In brief, the work has identified a major series of linear embankments, forming enclosures, throughout County Galway. A set of embankments appears to consist of 3 double embankments; an inner, a middle and an outer. The double embankments were constructed by digging a deep 'fosse' or drain and throwing the earth up on either side. The width of each embankment appears to have been about 60 feet. The set of embankments covers about 400 metres and the 3 embankments ran more or less parallel along the countryside. The route of the embankments was well chosen to include high ground into the embankments and thus reduce the effort of construction. In places, there were transverse embankments joining the inner and outer banks. These transverse embankments were single banks of 20 to 25 feet in width. There appears to be two centres to the enclosing embankments, one at Turoe/Knocknadala and one at Athenry. The set of embankments closest to the centre are termed the 'inner ward' and the embankments surrounding this are termed the 'outer ward'. Because of the close proximity of the two centres, the surrounding embankments collide with each other, making reconstruction difficult in these areas. Once the 'collision' occurred there appears to be additional sets of embankments surrounding the two centres and joining them into one very large enclosure. Remnants of the embankments are to be found throughout the County. In different areas they are called Dun Cladh, Muc Cladh, Mootes or double-ditches. While people would be familiar with the embankments in their own area, they would be unaware of similar embankments in other areas. This is the first time that the various vestiges have been linked together, and the results are very interesting.

These embankments would have been constructed over several centuries as the power of those in control expanded and their needs increased. For reasons elaborated later, the embankments appear to be defensive in nature. As the reconstruction proceeds it becomes obvious that the embankments were cleverly constructed. Bog, marsh, lakes, forest and high ground were incorporated into the embankments to make them more effective defensive features. In many places the banks of rivers were deliberately raised. This allowed floodgates to be fitted so that, when closed, the river level was raised, flooding large areas of low ground.

So, if these were defensive features, what was being defended? Again, as will be elaborated later, legend states that a Celtic Tribe called the Fir Belg landed at a port near Clarinbridge called Ath Cliath Magh Ri. Within the Fir Belg there were various tribes called the Manapi, Atrebates, Marassi, Fotads etc. Legend further states that they set up their Centre of Power at Cnoc na Dail. A typical Celtic Centre of Power would have an Assembly point, a Ceremonial site, a Royal cemetery and a place for Poets. All of these hallmarks of a Celtic Centre of Power are to be found at Turoe/Knocknadala. The linear embankments appear to be surrounding Turoe/Knocknadala and, therefore, defending it.
Further legend recalls Athenry as a Centre of Power and the linear embankments surrounding it corroborate this to some extent.

A more massive embankment was constructed by the main people of Ulster, who were called the Cruthin. This is known as 'Black Pigs Dyke' ('Claidh na Muice Duibhe') and is thought to have run from Donegal on the NW coast to the N of Dublin on the E coast.

This was always claimed to have been a defensive embankment to defend themselves from the expanding Connaught peoples. In many ways this embankment was similar to the present embankments. The Black Pig's Dyke, however, ran in a line from coast to coast. The Galway embankments, on the other hand, appear to form enclosures.

Throughout the book an interpretation/translation of the townland/placenames is given in brackets. As the linear embankments cover increasingly larger areas, the number of townlands also increases. While this appears to add to the complexity of the book, it is a fascinating insight into the features, events and people in legendary history from 2000 years ago and how these are recorded in many placenames.

The complete book on this subject is called **'Hand of History – Burden of Pseudo-history'**. In order to set the background, it covers in great detail the events in Europe and Britain around 50 B.C. It goes on to reconstruct the linear embankments in Galway in minute detail. Finally, a detailed and thoruogh literature review is used, in association with the current research, to show that the Turoe/Athenry area was at the centre of Irish history 2000 years ago. This also shows how early Irish manuscripts were modified and invented in order to supress the Turoe/Athenry Centres of Power.

The following short version of the book is intended for popular reading. The available information has been simplified as much as possible to make the book interesting and informative for the reader, but also to retain the completeness of this remarkable story of the importance of Turoe and Athenry in pre-historic times. The book does not contain references, but does have a list of relevant references for additional reading. The book **'Hand of History – Burden of Pseudohistory'**, which has been extensively researched in terms of documents, archaeology, the findings of more recent historians and oral history, does contain all the relevant references for those interested.

Overview

In the years leading up to 56 B.C. the Romans had conquered much of Celtic Europe. Some of the Northern Celtic tribes moved westwards, many of them fleeing Europe to Britain. When Britain was invaded by the Romans in 55 and 54 B.C., the Celtic tribes put up a determined fight. While they were eventually defeated in 54 B.C., the Romans were unable to consolidate the victory because of further uprising in Europe. While rumours of further invasions were rampant, many of the important Celtic leaders in Britain disappeared from the records. They certainly didn't flee into Roman Europe. At the same time as these leaders disappeared from British history, Irish legendary history begins to record a number of Celtic leaders who may well be those who disappeared from Britain. At that time there was a well-known sea-port called Ath Cliath Magh Ri near Clarinbridge in Galway Bay in the West of Ireland. It was through this port that the legendary Celtic leaders led their people. From there they made the short journey inland and set up their Centre of Power at a place now called Knocknadala, or Cnoc na Dail, meaning Hill of Parliament or Hill of the Great Assemblies. Just beside this they set up their Ceremonial site on Turoe Hill and placed the beautifully decorated Turoe Stone there. This area is 7 miles east of Athenry and 5 miles north of Loughrea, in County Galway. They had taken this territory from the native Cruthin people, who may have already had a Centre of Power at Turoe. Fearing a Roman invasion, and in an effort to defend the land they had won, they set about constructing a series of defensive linear embankments. They had already seen how successful the linear embankment defensive system was against the Romans in Britain. So a similar system was constructed to enclose Turoe/Knocknadala and defend it. As the area under the control of the Celtic Tribes expanded, so too did the

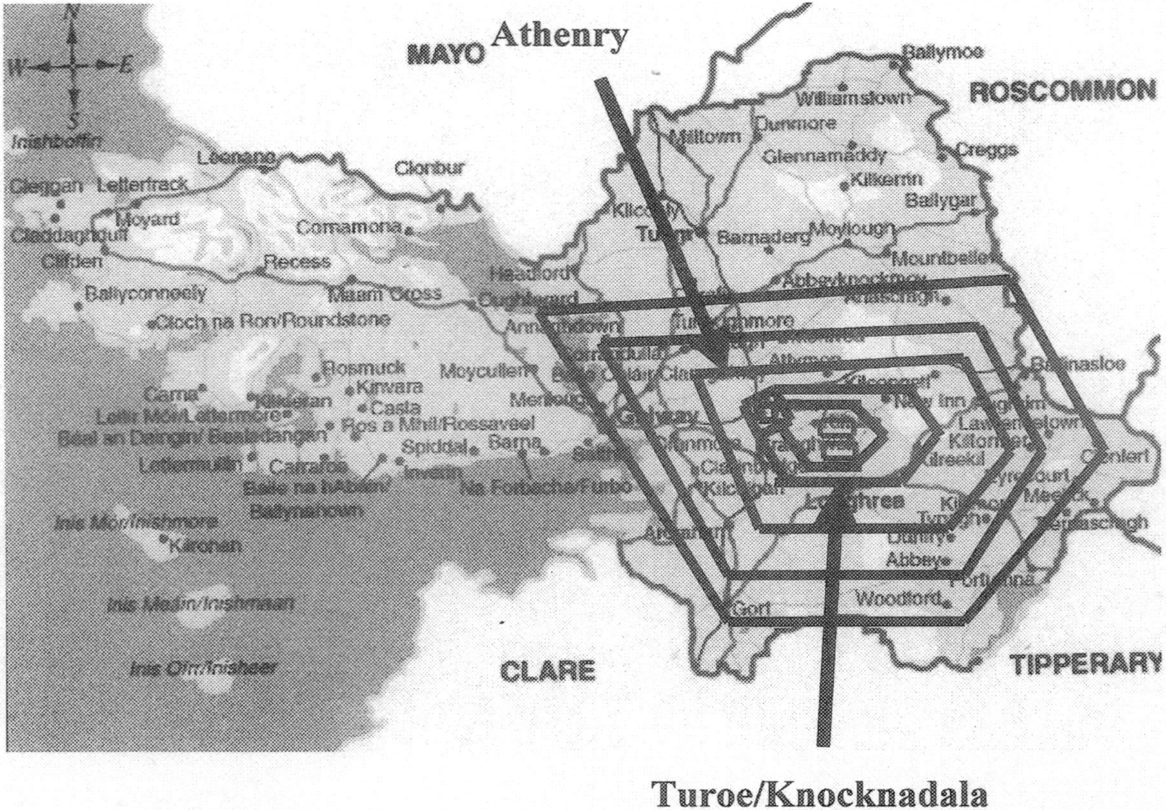

Map of Galway showing some of the linear embankments surrounding Turoe/Knocknadala and Athenry, and some of the further extensions surrounding both centres. For clarity, only a few of the outer ward extensions are included.

linear embankment defensive system. Further linear embankments were constructed to enclose a larger area. When a second Centre of Power was established at Athenry, linear embankments were constructed to defend it in similar fashion. The proximity of the two Centres of Power meant that they could not survive independently. For the next few centuries these two Centres of Power expanded as one. As land was won from the Cruthin, linear embankments were erected to defend it (see map of Galway with lines of some of the embankments indicated). Even though the threat of a Roman invasion had passed, the system of linear embankments had proved successful against the Cruthin and so was continued.

Irelands oldest legendary tales are called the Ulidian Tales and tell stories of constant warfare between the men of Ulster (the Cruthin) and the men of Connaughtmen. One of these tales is the well-known story of the Tain Bo Culaigne, where Queen Medb of Connaught tried to steal a bull from the Ulstermen.

The renowned Greek geographer, Ptolemy, left a record of the geography of Ireland as he knew it about the year 150 A.D. One significant feature of this record is that he knew of two Royal Capitals, one in Ulster and one in Connaught, in the vicinity where Turoe is today. He also recorded that the most illustrious city in the British Isles was in the south of Connaught.

Eventually, the Royal Centres of Turoe and Athenry were to fall from use. But worse still, in order to glorify the Kings of the 700's/900's A.D., the traditions of Turoe were transferred to Tara of Meath and the traditions of Athenry were transferred to Rath Croghan of Roscommon. This was reasonably easy to achieve, since Tara and Turoe both share the same name in Irish, Temhair. Athenry in the early days was called Rath Cruchain, a name easy enough to change to Rath Croghan.

For this reason, Turoe and Athenry are not mentioned in our current history books. However, the oral traditions of the area are very different from what is in the history books. This, along with the discovery of a complex series of linear embankments in Galway, an enormous number of significant placenames, the Ulidian Tales and Ptolemy's record certainly calls for a total re-evaluation of the history books with regard to the Ireland of 1500 to 2000 years ago.

The following chapters of the book give details of the linear embankments and the route they followed through the Galway countryside. In addition, the rich oral history and tradition of the Turoe/Athenry area, much of which is almost lost, is recounted. This can become complex as the townlands through which the linear embankments were constructed become more numerous. Quite frequently though, the townland names reflect some feature of the linear embankments or some historical person or event of the time. So, bear with the occasional complexity as a fascinating story of Ireland as it was 2000 years ago begins to unfold and the realisation dawns that Galway was the centre of the Irish developments at that time.

Location map for the area dealt with in Chapter 1. This area a few miles east of Athenry and north of Loughrea. (Map courtesy of Ordnance Survey Ireland, Permit number MP 0001903).

Figure 1.1. The Turoe Stone.

Chapter 1.

Linear embankments around Turoe & Knocknadala Hills in County Galway

In the past very little has been published on linear embankments in Galway. While the well known Black Pigs Dyke stretching across Ulster has been well documented and Martin Finnerty identified linear fortifications around Athenry in his book Punan Arsa, little else has been published relating to linear embankments in Galway. Yet, to many elderly people in various parts of County Galway (and elsewhere) linear embankments (referred to as double ditches, ramparts, mootes, dun cladh or muc cladh) are a well-known feature of the landscape. Some people would be familiar with these in their own area, or over a short distance, but nobody has ever tried to link up these linear embankments and relate them from one area to another, apart from Martin Finnerty's partial attempt in respect of the linear embankments surrounding Athenry. This Chapter of the book addresses this issue and reconstructs the linear embankments around Turoe (where the famous Turoe Stone is located; Figure 1.1) and Knocknadala townlands, about 7 miles east of Athenry and 5 miles north of Loughrea, in County Galway. As will become obvious from the size, extent and design of the embankments, they probably had a defensive role. Local knowledge of existing remnants of the embankments, the memories of the older generation on levelled embankments going back to the end of the 19th century, evidence from old Ordnance Survey (O.S.) maps and a field survey of the area have been used to achieve this.

Reconstruction of ramparts around Turoe Hill (Cnoc Temhro)

A set of 3 linear embankments encircled Turoe Hill. From the east they stretched the full length of Turoe Hill from Gortcam and

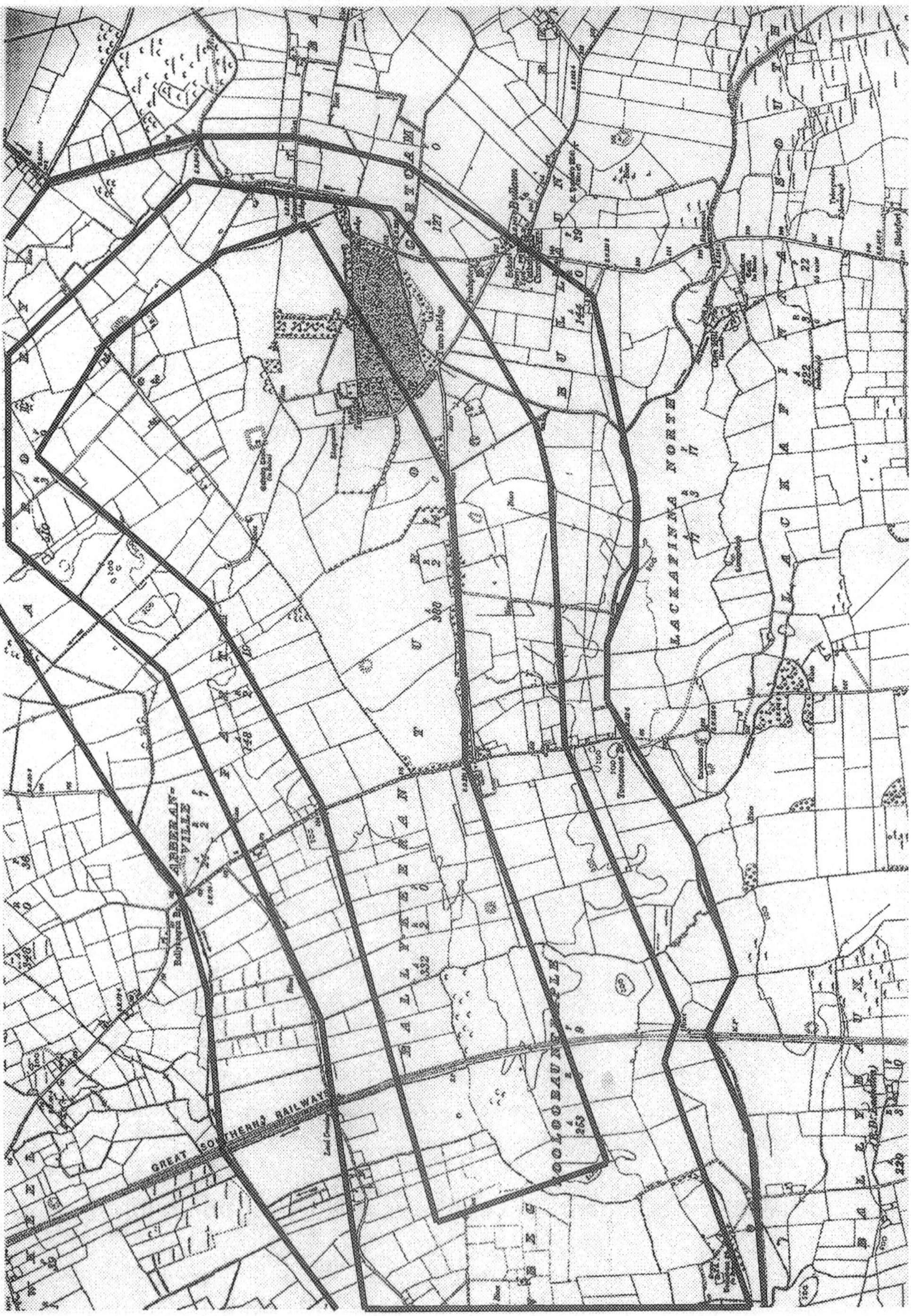

Figure 1.2. Map showing the triple set of linear embankments surrounding Turoe Hill. Reconstructed from surviving vestiges, old maps and sections that can be traced or are remembered locally. Drawn on the 1933 O.S. map. (Map courtesy of Ordnance Survey Ireland, Permit number MP 0001903).

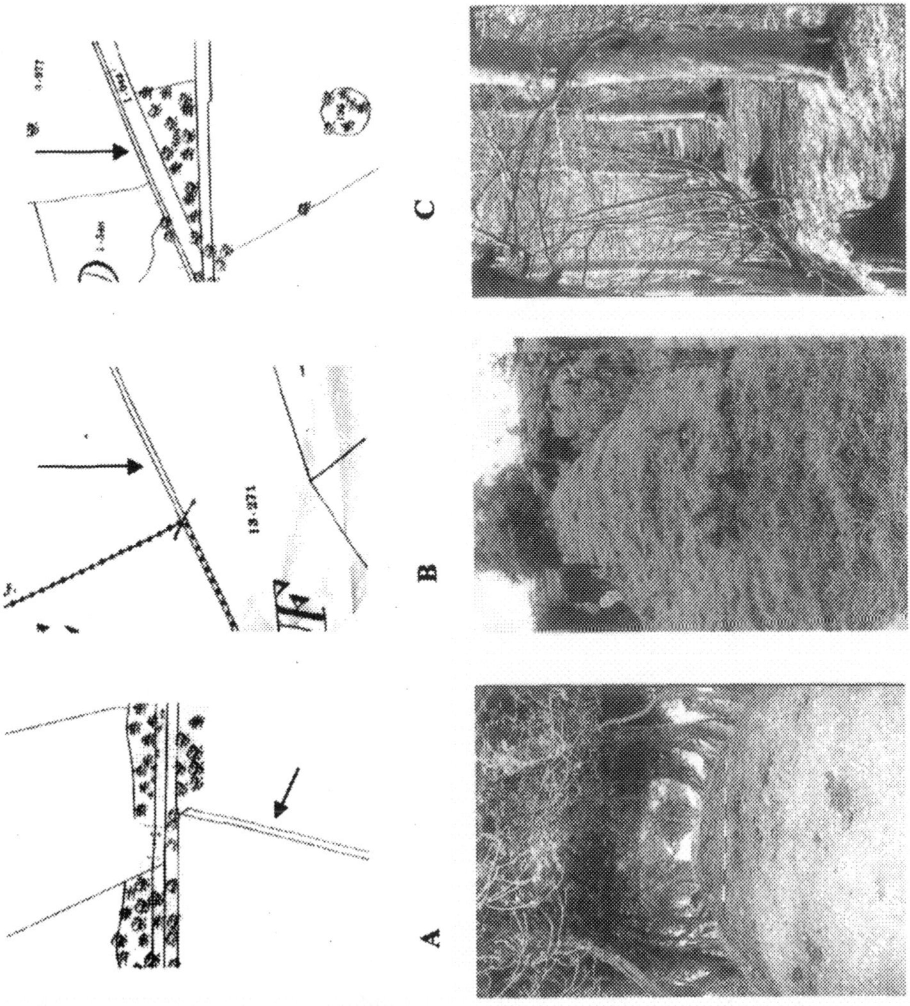

Figure 1.3. Methods used to depict the linear embankments on the 1894 O.S. 25" map (scale 1:1). Embankments are indicated by the arrows. A, a double line leading from the road; B, a double line; C, a triple line heading towards the road. Exactly the same methods were used to depict the embankments on the 1933 O.S. 6" map. What these features are like on the ground today are shown in the photographs. (Map courtesy of Trinity College Dublin).

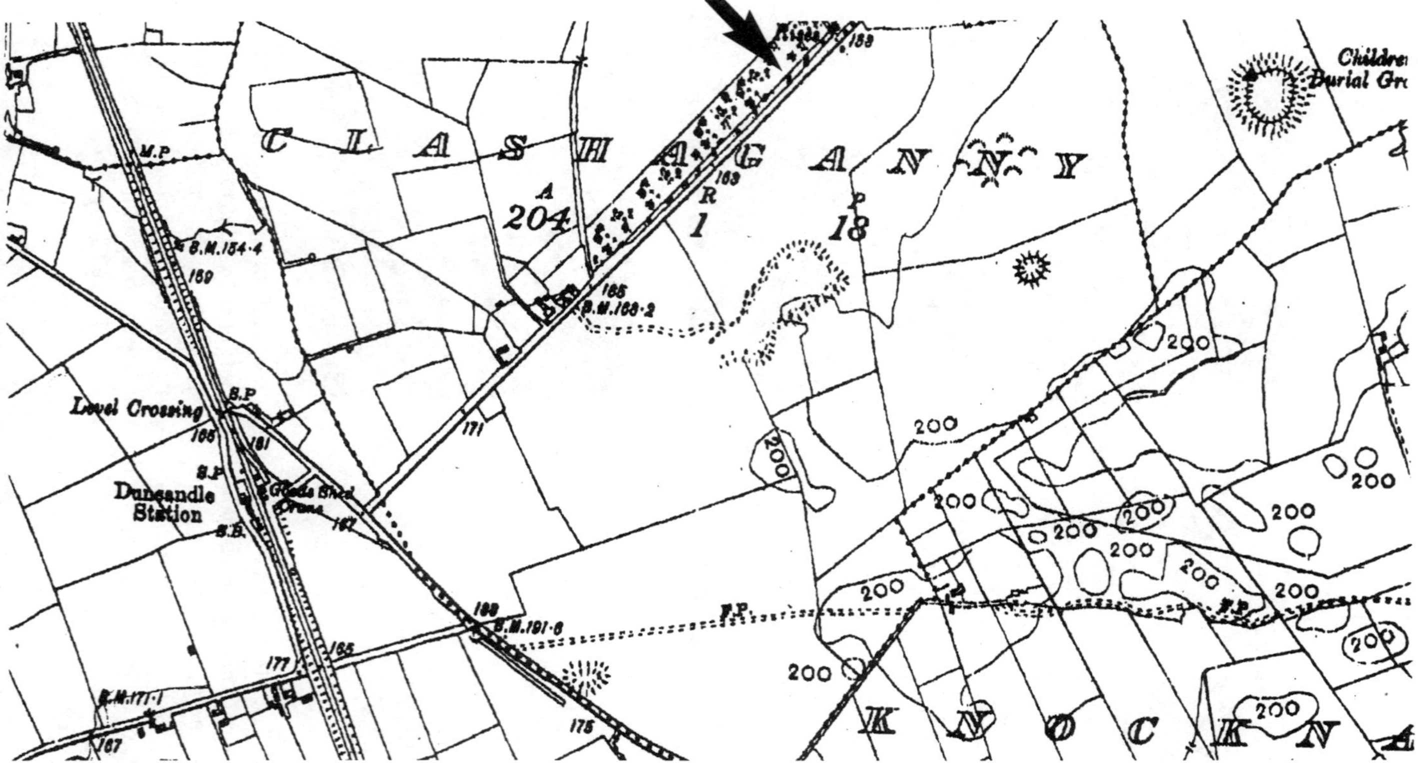

Figure 1.4. Depiction of the linear embankments, shown as a line, running alongside the Clashaganny road. As will be seen later, this feature of embankments running inside roads is very common. (Map courtesy of Ordnance Survey Ireland, Permit number MP 0001903)

Benmore along the heights of Turoe and Ballykeeran to Tooloobauntemple and Tooloobaunbeg in the West. From the north they stretched from Aberanville across the Hill to the boundary on the south of Turoe. These are shown on the accompanying map (Figure 1.2). Surviving vestiges of these embankments, except for those through the South of Tooloobauntemple are marked on the 1933 O.S. 6" map, Sheet No. 97. Similarly, they are marked on the 1894 O.S. 25" map (Figure 1.3). Just inside the wall by the road on the south of Turoe Hill there are remaining vestiges of the rampart (shown as a line on O.S. maps) that was levelled to build the road. Frank Corcoran, who lives on this stretch of road, remembers his grandfather recounting how this rampart was 14 feet high and 12 feet wide at the top in his young days. On the north side, vestiges still survive as shown on the 1894 and 1933 O.S. maps running ENE through the S of Fearta and the centre of Galboley to the SE boundary of Sraheendoo. Similarily, there are remaining vestiges on the east side of Turoe. Many other sections of levelled ramparts are still traceable. Indeed, Patsy Keane of Fearta remembers these ramparts surrounding Turoe and how he and his father worked hard to level many of them.

Reconstruction of ramparts around Knocknadala

A further set of 3 parallel linear embankments encircled the summit of the adjoining Knocnadala Hill. The road through Clashaganny on the N flank was built along the fosse of a levelled rampart (Figure 1.4). The road running S along the E foot of Knocknadala Hill through Carrowreagh and Galboley along the edge of Pump Bog was built on top of one of the inner ward ramparts which accounts for its height above ground level. These embankments surrounding are Knocknadala Hill shown in Figure 1.5. Many vestiges of these embankments can still be seen today (Figure 1.6), The embankments as they appear on the ground are shown in Figure 1.6.

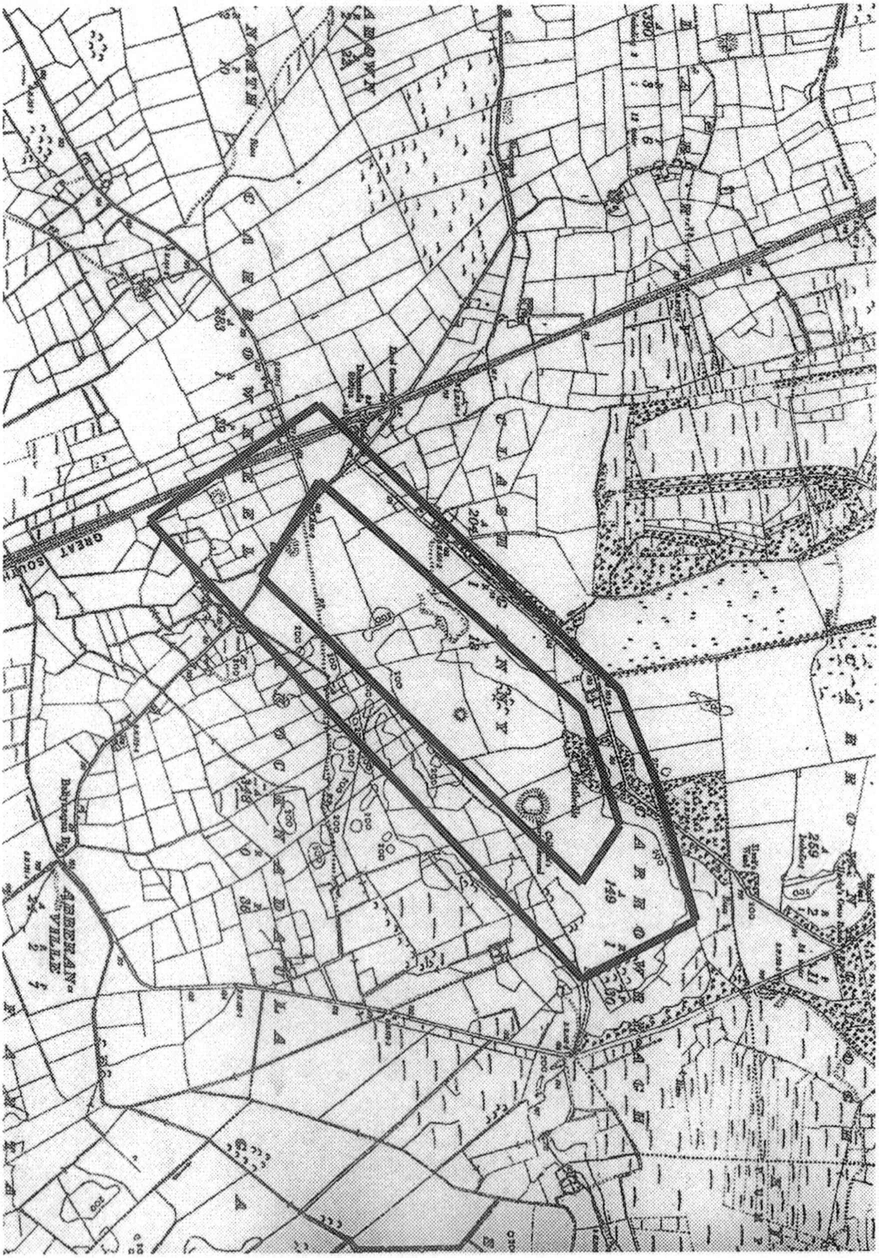

Figure 1.5. Map showing the triple set of linear embankments surrounding Knocknadala Hill. Reconstructed from surviving vestiges, old maps and sections that can be traced or are remembered locally. Drawn on the 1933 O.S. map. (Map courtesy of Ordnance Survey Ireland, Permit number MP 0001903).

Figure 1.6. Remnants of linear embankments as they can be seen on the ground on the N, S and E of Knocknadala Hill.

Figure 1.7. Map showing the full extent of the embankments surrounding Knocknadala Hill. (Map courtesy of Ordnance Survey Ireland. Permit number MP 0001903).

The original embankments, called the inner ward, were expanded W and NW to enclose a growing urban complex sprawling across Carrowkeel and Clogharevaun to the Kiltulla North boundary. The final expansion of the inner ward stretched all the way from the W boundary of Killescragh in Pump Bog, across Knocknadala Hill to the W side of the Bookeen road in Carrowroe North, Clogharevaun and Kiltulla, as shown on the accompanying map (Figure 1.7). This significant extension to the original Knocknadala inner ward included the ancient urban complex of Cotni (see Figure 1.8) and indicates a sizable population increase. Cotni (Cetni) is a local name for a 40-acre area in Carrowkeel. The National Monuments and Historic Properties has placed a Preservation Order on Cetni/Cotni and listed it as a National Monument entered in the Sites and Monuments Register. There it is recorded as a "Complex of earthworks in the townland of Carrowkeel, Parish of Kiltulla, Co. Galway, O.S. 6" Sheet 97." It is all that remains of an extensive urban-like population centre, which sprawled across Carrowkeel, Clogharevaun and Kiltulla. Prior to the Preservation Order much of the urban nucleus had been destroyed, carted away for limekilns or road building. In MacNeill's 'St Patrick', one population centre said to have been visited by the Saint was Campus Cetni in South Connaught, a place never identified. The renowned 2nd century geographer and cartographer, Ptolemy of Alexandria in Egypt, recorded a dense population centre in the west of Ireland as *"the most extensive and illustrious acropolis (urban population centre) in all Britannia."* He called this Nagnata(l), which easily translates from Greek to Cnoc na Dail or Knocknadala, a townland next to Carrowkeel. Corroborating Ptolemy's astounding record, elderly locals claim that several segments of the W, NW and S flanks of the Turoe/ Knocknadala complex, particularly Cetni (Cotni) were traditionally regarded as part of an ancient town or urban centre. They pointed out ancient "streets" of little houses, close-set ringforts and a complex of earthworks. As further corroboration of Ptolemy's record and local tradition, early Dindsenchas (history of royal sites) material referred to *"the dense population"* of this area: *"O Maen na treibh tuillte co rein na fairrigi."* meaning "from densely populated Maen Magh to the sea." Mean

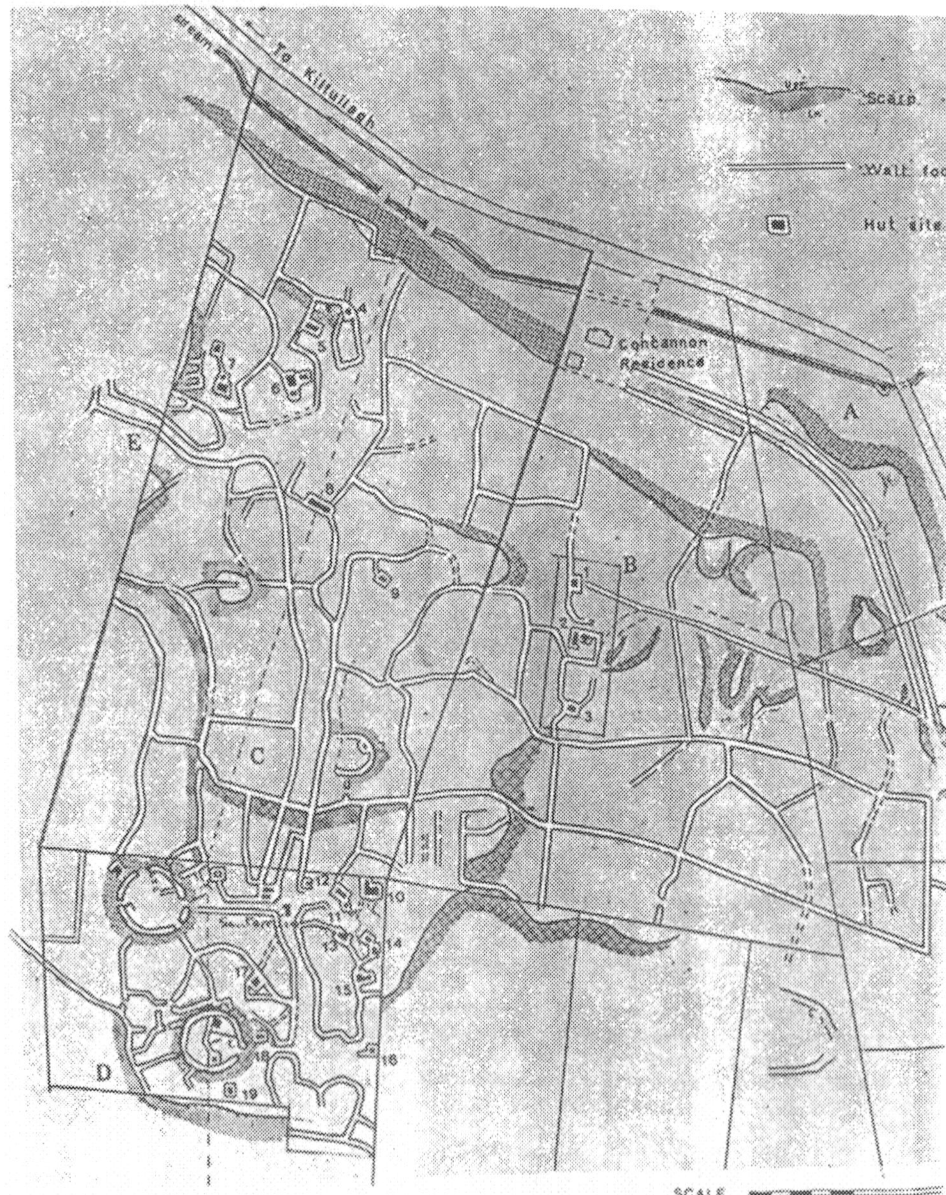

Figure 1.8. Detailed map of Cotni (National Monument in Carrowkeel) showing field system, house bases and ringforts. Reproduced with the permission of the National Monuments & Architectural Protection Division.

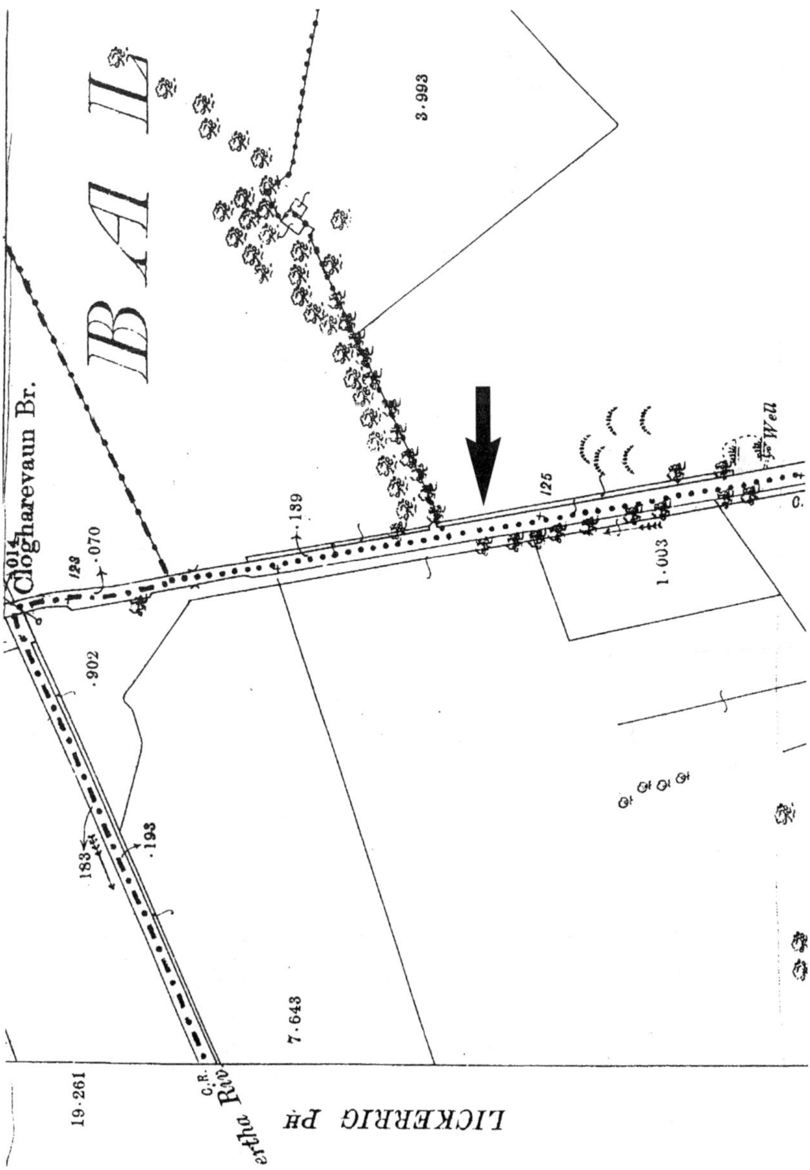

Figure 1.9. Depiction of the linear embankments, shown as a line running alongside the road, on the Bookeen road. (Map courtesy of Trinity College Dublin).

Magh is the ancient name of the plain at whose centre stands the Turoe/Knocknadala complex. Significantly no other area of ancient Ireland was accredited with a dense population as was Maen Magh. The current thinking that the Irish in pre-history did not have urban centres is in direct conflict with this striking evidence.

The 'Dindsenchas' are a set of ancient documents that tell the legendary history of places in Ireland. In the 10th/11th Centuries the oral traditions of those places were transcribed and recorded. In many instances the information recorded in the Dindsenchas records is at variance with 'official' history. Hence, historians have viewed many of these documents with scepticism. However, the frequency with which the Dindsenchas of places around Galway are corroborated by archaeological evidence, current oral history and placenames is astounding. This, allied to the evidence that many early manuscripts e.g. The Book of Armagh and The Book of Leinster, were deliberately modified calls for a total re-evaluation of the Dindsenchas records.

The late Mrs. Smith and Ml. Donnellan of Ballyknock walked to school in the latter 19th century along the "big double ditch" through the E side of Bookeen South and North to Ballynahown. This continued N along the E bank of the river on the E boundary of Clogharevaun and W boundary of Killarriv. Surface traces of transverse ramparts from this inner embankment of the set to its outer counterpart survive along the Bookeen North/South boundary running into Carrowroe North, along the bank of the river running W from Clogharevaun Bridge, and another across the centre of Clogharevaun along its highest point W of the road to the Kiltulla South boundary. Vestiges of Ranpart along the Booken Road are shown on the 1894 O.S. map (Figure 1.9). From the edge of Kiltulla cemetery the set swung E behind Kiltulla church following the swing of the Raford River which it straddled as it ran E through the N side of Killarriv (Figure 1.10). In the early 20th century, now deceased members of the Ford, Kennedy and Ward families pointed out vestiges of this set of embankments through

Figure 1.10. Various sections of stripped down ramparts in Killarriv. Left, close to the river at Kiltullagh Bridge and right, in Killarive bog. Locally, this bog is called Poll 'a Chladh (bog of the dyke rampart).

Killarriv, naming them Curragh More, Curragh Beg, and at the foot of Killarriv hill simply as "the curraghs". All were levelled more than 60 years ago. The banks of the Raford River just NE of Kiltulla Bridge and again E and W of the two Raford Bridges were built up and sluice-gates fitted to create artificial lakes to enhance the defences around much of the N and E flanks of the Turoe/Knocknadala inner ward. The area of Killarriv Bog through which they passed is still remembered as Poll 'a Cladh (Dyke Bog). This extended inner ward set joined the original set of ramparts in Carrownaglogh (Carrow na gChladh = Dyke Quarter) and ran SSE through Pump Bog to Sraheendoo. Following the stream, which is the boundary between Knocknadala and Turoe, the set of linear embankments ran W through the S flanks of Knocknadala, Carrowkeel and Bookeen South.

This set of ramparts enclosed the townland of Cloghereavaun (Clogher Ui Mhain). Deceased members the local Kelly, Murphy, Mullins, Duane, Ford and Kavanagh families claimed the whole area from Cloghereavaun to Kiltulla cemetery area was a continuation of the urban nucleus of Cotni (Campus Cetni in Carrowkeel) across the river. These areas were linked by an esker roadway, which was recently demolished in gravel mining. In the 1800's, when Landlords vied for development projects, Dunsandle and Kiltulla Houses conspired to have a proposed railway line from Monivea to Loughrea pass close-by to serve both. The proposed line was to pass through Cloghereavaun and an area within that townland known locally as Satchell. It is not known why this local name exists, but it is interesting that among the obstacles demolished in a rush to provide the best route for the railway line were a rectangular site, Downahachell *(Domnach Shachell?)*, an adjoining cloister or cemetery, Kelhachell *(Cell Shachell)* and a fort known as Rathatchell *(Rath Shatchell)*. *'Domhnach'* denotes an early church and *'Cell'* refers to a monastic cloister or cemetery. The attached name, Sachell, is that of the founder. The sole monastic founder of that name was the pre-Patrician Bishop, Sachellus, who led missionaries from Gaul to Ireland in the train of Palladius about 431. Sachellus may well have established his original church beside the Turoe/Knocknadala Royal Palace, where he would have had Royal support and a population centre. Indeed, this is corroborated by the fact that ancient records and maps refer to a very early (pre-7th century) monastery in the Kiltulla/Clogharevaun area. Eventually the idea of a railway line here was aborted in favour of the Attymon/Loughrea line. This area in the S of Clogharevaun is still known locally as Sachell.

Joint Turoe/Knocknadala oppidum inner ward

Parallel linear embankments were constructed across the stream and hollow between Turoe and Knocknadala hills linking their seperate enclosing linear embankments on their E and W flanks to fuse the two Hills into one joint inner ward, one archaeological site. Vestiges of those link embankments are shown on the 1933 O.S. 6" map, Sheet No. 97.

Reconstruction of theTuroe/Knocknadala oppidum outer ward ramparts

The linear embankments just described constitute the inner ward of the Oppidum, Population Centre and Centre of Power. Surrounding this inner ward were another set of embankments, the outer ward. Working clockwise from the SE corner, the set of three linear embankments ran W from Oosigh Bog (*Poll na nDhoosigh* = Rampart Bog) in Bellayarha South (a mile SE of Turoe) through Lackafinna South, Lackalea (*Leac a' Chlaidh* = Dyke Hill), skirting Tobergradan and Tobernalugh passing Tober Breedge, through Cahernamuck and Ballybaun and into Caherhenryhoe. A string of sentinel forts stand on the hilltops along this line of defence. Among these are Lisnagalagh and Tonnawausa (*an tShonnagh Mhasa* = Broad Hill Rampart). Caherhenryhoe's broad hill was the focus of a dense settlement area including several cahers (stone forts) bearing the names of ancient kings. Elderly locals now departed, such as Pat Curley, Pat Daly, Ml. Moran, Martin Quinn and Ned and Michael Ryan, to mention but a few, claimed that elderly people in their young days spoke of a Cahertoohal, Caherquinn or Cahercon and a Caher Cormaic along this line of now obliterated massive stone forts with several large souterrains. The Royal connotations of these forts, which have the names of alleged 'Ri Temhro' Kings, along with the closeby forts of Tinny (Cahertinny), Tuathal (Caher Tuathal), Maine (Cloghereavaun) and Na Thi (Cloghernahy), alleged Kings of Connaught, are significant. This corroborates the unique name of Caherhenryhoe townland, translated as Caher na Ri Uabha or Royal Cemetery of the Forts of the Kings. This, along with the evidence to be presented later, certainly makes a case for Galway being the true 'Home of the Kings'.

These old men described the unique character of Caherhenryhoe hill (demolished in gravel mining operations over the last 60 years). In the early 20th century, soldiers re-erected the fallen 14ft high standing stone on the summit of this formerly cone-shaped hill. This was surrounded by a circle of granite stones, one of which was a 'bullaun'. Some 'cairns' (stone burial mounds) called *Uabha,* which constituted part of the Townland name, stood some distance back from this central megalithic setting. All of these collapsed as they were undercut in gravel mining. The 'bullaun' and some of the granite stones were pushed to nearby walls, while the tall 14ft standing stone lies broken in two at the bottom of the gravel pit. A line of granite standing stones were said to have run from here to Lisnaghalagh (Fort of the Standing Stone) on Tonnawausa hill near Turoe. Just across the river beside the ancient cemetery mound on the top of which stands the ancient tiny Tooloobauntemple church is a large undecorated Turoe-Stone-like granite stone. An almost lost local tradition claims it marked the burial mound of King Tuathal which gave its name to this Townland, Tooloobaun = *Tuathal Ubhan* = the Burial Mound of King Tuathal, the grandfather of King Conn who gave his name to Connaught.

Along this S flank, as elsewhere, the field grid was dictated by the rampart alignment. From Caherhenryhoe the E/W field-grid alignment dictated by the linear embankments along the S flank took on a NNW alignment in tandem with the rampart set as it swung NNW through the E part of Carrowbaun, Dunsandle, the centre of Lisnadrisha, and straddled the Raherneen/Cahernalee (Caher na Chlaidh = Dyke Fort) boundary. Land reclamation demolished the ramparts along that boundary up to the edge of the former lake on the Dooyertha (Dughiortach/*Dur*) river, now known locally as 'Ceallagh Bui'. It was a lake before the new river-channel was cut. Floodgates further W served to flood an extensive area for defensive purposes.

This set of ramparts continued to Kiltulla South and its border area with Greyford where there are massive stone walls some 10 feet wide. Local tradition claims the whole area of Kiltulla South between the high, formerly wide stone-wall ramparts of the two seperate defensive lines bounding the townland was maintained as a deer-reserve from the late middle ages until the early part of the 19th century. Indeed this can be seen on the 1817 map. High walls built of stone from the massive ramparts sealed off the S and N ends. Deer were kept in this enclave by local landlords for the regular hunt through the countryside. When the victorious English army was on its way to Galway after the Battle of Aughrim they camped on the lawn of D'Arcys House. All the deer were killed for a feast and the reserve never regained its former glory. Thousands of tons of stone were carted away from these massive walls to lime-kilns and road-making operations of the 19th and 20th Centuries. The old massive walls were well known to old men of the vicinity who in their young days had ripped the stones apart hunting rabbits and other wild animals taking refuge there. The last remnant of the massive walls was carted away to the stone crushers of more recent times.

From Greyford (Rath Chlaidh = Dyke Fort) the set straddled the W boundary of Kiltulla North up along the W edge of the bog before swinging sharply E, slightly ENE, straddling the bog and the

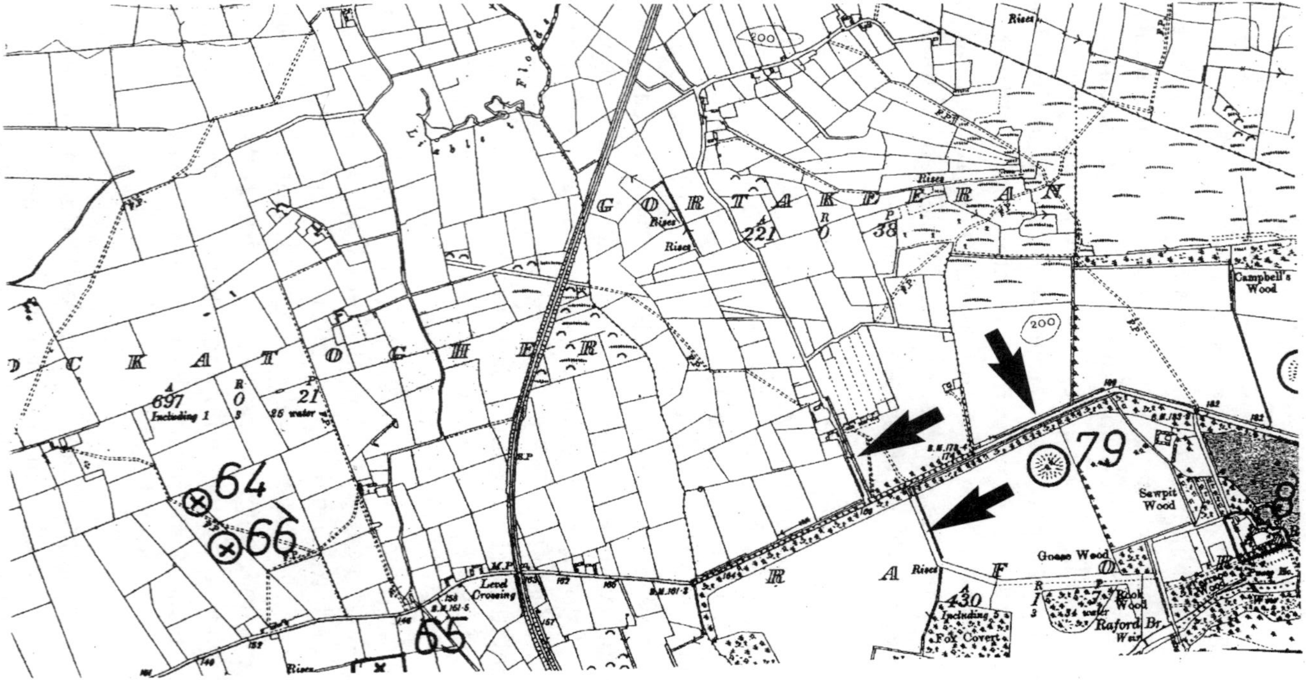

Figure 1.11. Ramparts in the Raford area as shown on the 1933 O.S. map. This area is locally knows as 'The Rooauns', corrupt plural of 'roo' meaning rampart. (Map courtesy of Ordnance Survey Ireland, Permit number MP 0001903).

boundary with Carnakelly South. Several vestiges of rampart survive in Kiltulla North and are shown on the O.S. 6" map, Sheet No. 97, including an excellent section of a transverse rampart. The set ran on through Knockatogher and Gortakeeran, straddling the Raford road and the Raford/Knockroe boundary. Several vestiges of linear and transverse ramparts survived in Gortakeeran and Raford, such as that alongside the Raford/Gortakeeran road. Some are shown on the 1933 O.S. map, No. 97 (Figure 1.11), but many, inexplicably, were not shown. Numerous segments of the embankments along the Raford road were levelled in living memory. The area along that section of Raford is still known as 'The Rooans'. 'Roo' is Old Irish for 'earthen rampart', 'rooans' the corrupt plural form. From the built up banks of the Raford River forming the N boundary of Brackloon and the W side of Cloonsheecahill Bog the set of ramparts swung sharply SE heading to the W side of Carra More bog through Killescragh, Brackloonbeg and Benmore before swinging SSW through Bellayarha North and South back to the starting point.

A transverse rampart in Brackloonbeg long utilized as a right-of-way is known as 'The Green Road'. The dangerously high narrow road through Oosigh Bog in Bellayarha South was built on top of a rampart of this set. A large fort stands sentinel beside a rampart of the defence line through Killescragh and the ancient Escir Riada highway crossing it at that point. Lisheenahassy (*Lis na Shassaigh* = Fort of the Embankment) inside the W boundary of Carramore heralds the rampart route.

One of the linear embankments through Brackloonbeg and Killescragh was first levelled down as the original route for the Loughrea/Attymon railway line. It is still known locally as 'The New Line". When a different route was later found for the railway line, this levelled section through Brackloonbeg/Killescragh was built as the present road. In the northern half of Brackloon the same linear embankment of this outer ward was similarly cut down and laid out for the railway line and was similarly known as 'The New Line' (Figure 1.12). However, after that railway line

Figure 1.12. A stretch of levelled rampart in the north of Brackloon. Called the 'New Line', this was prepared for use as a railway, but was never completed. It remains as a 20 metre wide track with embankments on each side (see photo insert). (Map courtesy of Ordnance Survey Ireland, Permit number MP 0001903).

route was rejected, this segment in Brackloon remains exactly as it was left without ever being developed into a road. It is an excellent example of how these ancient linear embankments were being utilized with part of the inner bank and outer lip rampart being retained on either side. As a boy Bill Joe Craven and his schoolmates from Shanvalla in Cloonsheecahill walked to school along this stretch of embankment. He pointed out the route of this set of linear embankments all the way to Bellayarha, a distance of 5 miles, and a similar distance back in the other direction, a feat very few others could do. At most, they know of a 6 or 7 miles stretch of rampart after which they have no idea of its onward course. Formerly, horse carts used these ramparts as right-of-ways causing considerable deterioration. Laws were passed forbidding carts, allowing use only by horse and rider or on foot.

Another excellent example of this modern employment of the ancient ramparts is the section of road which forms the southern Brackloonbeg/Benmore boundary with wide embankments of this set still in place on either side of the road (See Figure 7.10). Numerous stretches of the stream from Turoe Bridge to Carrowmore Bog were built up with floodgates fitted at different stages to flood extensive areas along this flank and enhance the defenses in times of danger.

Remembering that one line is used to depict the set of three linear embankments, this set of outer ward ramparts are shown in Figure 1.13.

First extension of the Turoe/Knocknadala oppidum outer ward

An extension of the Turoe/Knocknadala oppidum created a new outer ward. Starting once more from the SE corner at the Kilboght/Drought boundary, the set of linear embankments ran W through the N of Kinkullia straddling the Curraghroo/Coorbaun boundary with Bushfield. It ran astride the road through Lackalea, Cahernamuck East and West, Cahernamona and Raruddy East and West. Across the river from Lynchfort Castle, the stream was

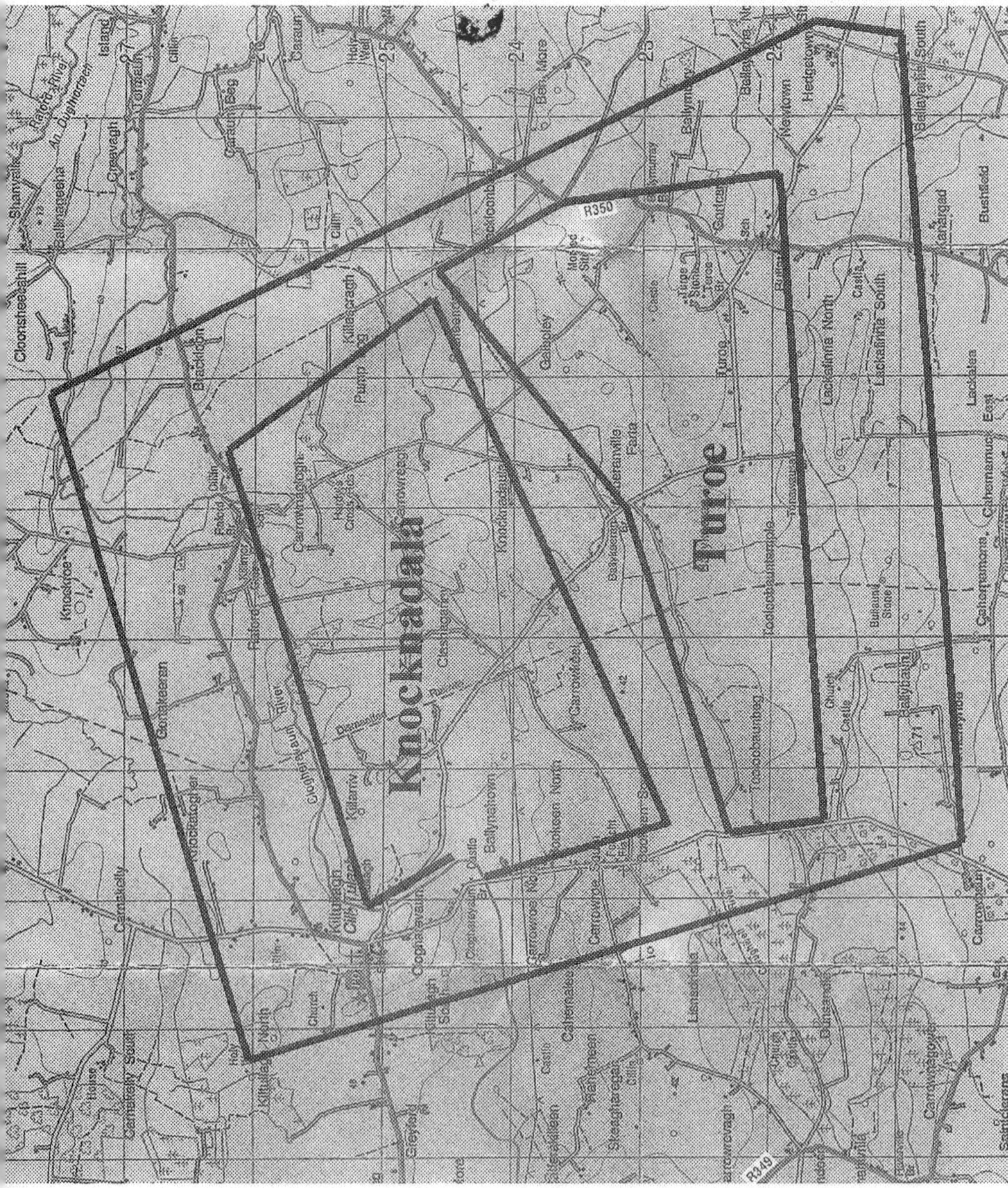

Figure 1.13. Rough position of the first extension of linear embankments surrounding Turoe and Knocknadala Hills. For the purposes of clarity, the set of three embankments are represented by one line. (Map courtesy of Ordnance Survey Ireland, Permit number MP 0001903).

banked up and floodgates fitted to enhance the defences by flooding in time of danger. This system is still operable and has been used for land-fertilization purposes. The ramparts ran into Cahertinny and Doogeraun, skirting St. Faelan's Well, and Killilan Bridge (Cill Fhaelan), named after 7th century St. Faelan who had his monastery here. The rampart set raced W through the marshlands of the St. Cleran's River whose banks were built in places on the Cahernaman/Cahertinny boundary with Caherhenryhoe as shown on the O.S. map. From its boundary with Doogeraun (Dyke Compound) the set swung NNW along the Carrownagower/Carrowbaun boundary through the W side of Dunsandle where vestiges of the ramparts survive in the strips of woodland. It then straddled the Lisnadrisha/Carrowrevagh, Skeagharegan/Carrownamorrissy, Caherakilleen/Oldcastle and the Curraghmore/Rathgorgin boundaries. In the 1840's long stretches of a linear embankment through these townlands were levelled for the laying of the proposed Limerick/Loughrea/ Monivea/Sligo railway line, still remembered as 'The New Line'. The great Famine ended this project.

Two close-set odd-shaped fortlets facing one another stand on either bank of the original Dughiortach (Dunkellin) River channel which funnelled here to a very narrow point between the two. They played a strategic role in controlling floodgates set on the river between these fortlets. They were backed up by raised banks along the old channel and stout outer ramparts to enhance the flooding of extensive areas along the river. Did they perhaps control river transport? The river-edge fortlets stood on the line of the set of ramparts along the W boundary of Greyford. A linear embankment forming the boundary survives as a long stretch of impenatrable thornbush-covered embankment, a sample of what the top of the original ramparts may have looked like (See Figure 7.7).

This set of linear embankments straddling the Curragh More/ Greyford (Rath Chlaidh = fort of the rampart) boundary marched with the outer ward linear embankments of the Athenry oppidum

Figure 1.14. The many vestiges of ramparts along the road dividing Carnakelly North and South, as shown on the 1894 O.S. map. Sadly, few of these embankments remain today. (Map courtesy of Trinity College Dublin).

(see later) along this flank. The earthen rampart skirting Curragh More fort and running up to the Kiltulla/Athenry road was levelled for the continuation of the Brusk road which was never completed. It straddled the Brusk (*Bru A Scail* = Otherworld Abode of the god A Scail or Cian) boundary with Curragh Beg and Escir, skirting the cone-shaped Cruach Chian (now demolished in gravel mining) and Dominick's hill, the E side of Esker and Loughan Escir. In Brusk, right on the line of ramparts and close to a series of Fulachta Fia on the S slopes of Dominic's Hill, is a large raised square military-like encampment, marked on the O.S. map as 'a moated site'. The embankments straddled the Carnakelly/Gloves' (*na gCladhbhe* = Defensive Dykes) South boundary. From the E boundary of Gloves South and Gloves Middle this set of the Turoe/Knocknadala outer ward ramparts swung sharply ENE straddling the Carnakelly road, which forms the Carnakelly North/South boundary (Figure 1.14). Part of this road was built on top of a rampart, which accounts for its height above the field level. The linear embankments of the Athenry oppidum outer ward continued N straddling the Gloves Middle/ Gloves East/Carnakelly North boundary.

Rampart vestiges survive where the ramparts ran ENE through Carnakelly North, Laragh, Lisduff (Figure 1.15), Killimor and its boundaries with Cappaghnanool and Lenamore. It ran along the Gannow/Derrynamanagh boundary to the Beech Hill/ Streamsford boundary. Ned Burke of Laragh recounted the route of the ramparts back through Carnakelly, Brusk, SSE to Dunsandle and Carrowbaun, and forward via Laragh Hill, Lisduff, Killimor and Lenamore to Beech Hill. Numerous remnants of ramparts have survived through Laragh, Lisduff and along the S edge of Killimor bog. The set straddled the Lenamore/Killimor boundary along the NW swing of the Raford river, which dictated its course. It swung SE along the Derrynamanagh side of the river to its boundary with Beech Hill. Marshlands along the river loop around the edge of Derrynamanagh bog were exploited fully to enhance the defences. The river and its floodlands backed up by ramparts formed a formidable defensive obstacle. The set ran

Figure 1.15. A gap at the corner of a rampart in Lisduff, where it swung south along the west side of Knockroe.

through Gortnaboha, Caraun, Knocknaboley and Bellafa, skirting Grange (*Grainseach;* corn repositories were located at strategic points).

A vast stadium-like embanked enclosure in Knocknaboley just N of the Bellafa/Ballinasloe road on the route of the Escir Riada occupied the flat area between Grange and Knocknaboley hills. The name Knock na Boley may spell out its purpose. Was this enclosure a cattle kraal? It was a universal practice to drive cattle into an enclosure at nightfall: "For it is the law of the Feini...that cattle should be in an enclosure (*i mbuaile*) by nightfall". Too large for such a purpose, did it serve a purpose similar to that of the Dorsey enclosure protecting the Sli Midhluacra entrance to Ulster's Capital of Emain Macha? This elaborate enclosure right on the route of the Escir Riada at Bellafa (Bel an Feadh = Forest gateway) may herald the strategic importance of this Bellafa gateway to the Turoe/Knocknadala oppidum through its forest periphery. This is further highlighted by the all-but-lost local traditions of Fianna camps in the vicinity. Townlands surrounding the Bellafa passage of the Escir Riada, through the forest periphery bear the names of 3 Fir Belg septs of the Fianna fighting men. These are the Clann Ui Conain (Mac Morna of Cloonyconaun), the Mac Rin of Crossmacrin and the Mac Lane branch of Turmacleane on the N slopes of Rahally above the ancient highway. This magnificent enclosure was recently bulldozed into oblivion to make way for the Third Millennium!!

The set of ramparts traversed Caraun More, Cross and Cloonyconaun and the NE slopes of Toormaclean and Rahally, straddling Brackloon and Slievedotia. On the Cross/Benbeg boundary near the route of the Escir Riada there is a Toberdoney (Well of the ancient church). Close by there was a site named Donaherneen. Doney/Dona seem to be the same Old Irish word *Domhnach* which indicates a very early church. The attached '(h)erneen' would appear to be the name of the founder. One of the early missionaries under the leadership of Sachellus of Donahachell (*Domnach Shachell*) in Clogharevaun in the NW

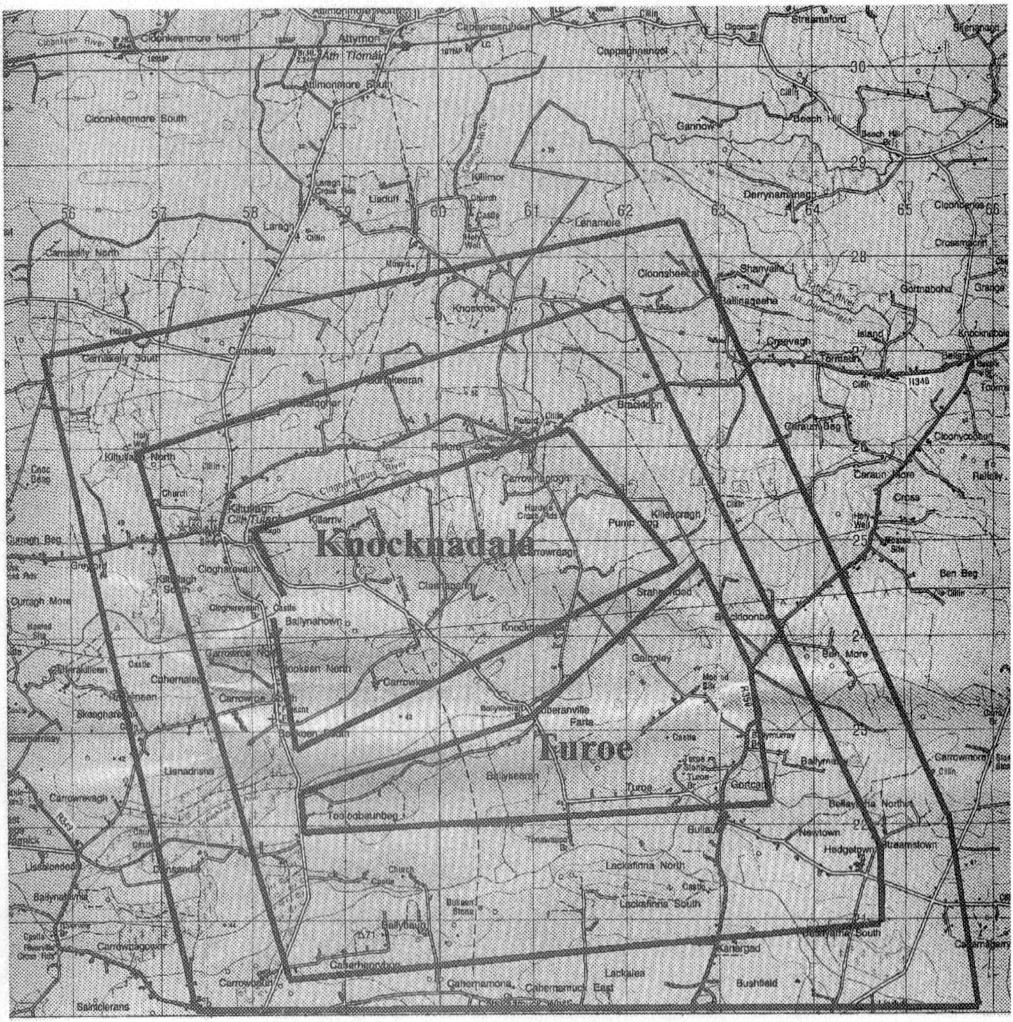

Figure 1.16. Rough position of the extensions of linear embankments surrounding Turoe and Knocknadala Hills. For the purposes of clarity, the set of three embankments are represented by one line. (Map courtesy of Ordnance Survey Ireland, Permit number MP 0001903).

corner of the Turoe/Knocknadala inner ward was Hernicus. What Donaherneen's relationship to Cross (*Crois*) townland and the nearby Turoe/Knocknadala oppidum core is not known. A short distance E of the holy well the large square military-like fort is recorded on the O.S. map as a 'moated site'. It is right on the line of the linear embankments, a regular occurrence along the defence embankments of this oppidum. Between this site and the holy well is another smaller site not marked on the map. According to an old man in the 1960's this was the site of Donaherneen. The rampart set skirted the Boleymore/Ballinphuill Bog to the Rathglass/Carra boundary where it swung S, slightly SSW, along the road into Carra and its boundary with Carrowmore. Excellent vestiges of ramparts survive all along the area. The banks of the stream running E/W through Carra had built-up banks to enhance the defences. Another stream from Carra bog forming the Doon/Cahernagarry boundary also had floodgates and raised banks to inundate a large area, including Carra bog, to enhance the defences in fear of invasion. The set of linear embankments ran on through Cahernagarry and Kilboght back to the starting point. Ramparts skirted the boglands and forested periphery along the N and E flanks just as they did on the S and W flanks.

This extension of the outer ward is shown in Figure 1.16.

Further extensions of the Turoe/Knocknadala oppidum outer ward

Further extensions of the Turoe/Knocknadala outer ward can be traced in a similar fashion, but at this point it is best discuss other features of interest in the Turoe area.

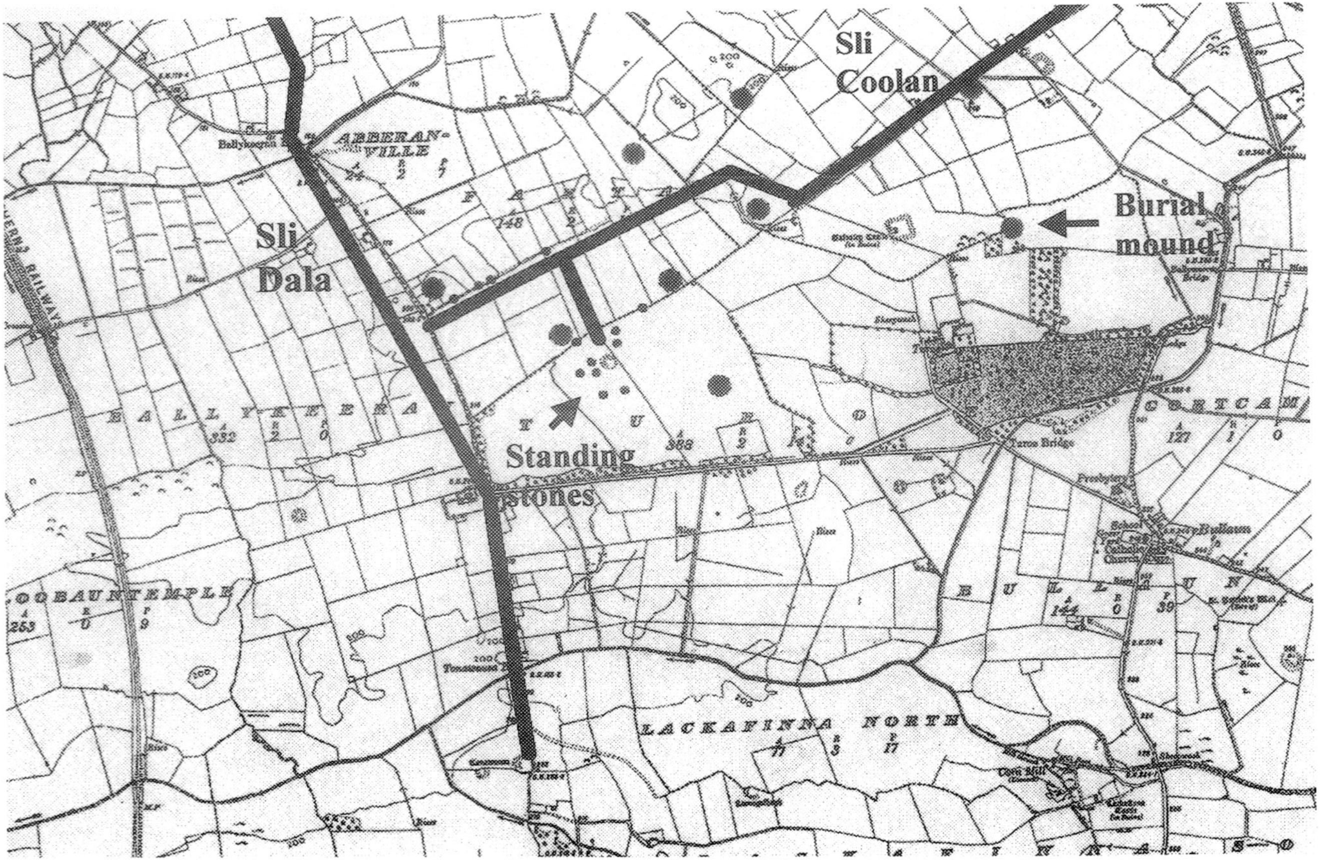

Figure 2.1. Map of the Turoe area showing ancient roads (black lines), standing stones (small black dots) and burial mounds (large black dots). Map courtesy of Ordnance Survey Ireland, Permit number MP 0001903.

Chapter 2.

Other features of interest in the Turoe area

Returning once again to Turoe/Knocknadala and the area enclosed by the first set of linear embankments, the inner ward. This area is 7 miles east of Athenry and 5 miles north of Loughrea. In 1914 and again in 1940 the Rath Feerwore (in Turoe townland), the ringfort where the Turoe Stone was originally located, was excavated. It was dated to 50 A.D. (although this date is now disputed by some), about the time that legendary history tells us that Dela led the Fir Belg invasion of Ireland through *Ath Clee Magh Ri*, the renowned Iron Age Seaport near Clarenbridge and Maree, in Galway Bay. During both excavations no locals were consulted about other features of interest in the area that might relate to the ringfort. If they had, they would have discovered a whole series of interesting sites, including burial mounds, ancient roads and the linear embankments (see Figure 2.1).

Clery's Hill

The highest point on the western half of Turoe near the original site of the Turoe Stone and Feerwore (or Rath Ferach Mhor, as it has been called locally), and overlooking a former expansive stone circle, there is a stepped hill mound known locally as Clery's hill (Figure 2.2). It is a veritable ancient cemetery in its own right. On its summit are what are referred to as 'sunken graves', having sunk visibly in living memory. Human-like bones have been knocked out from the edges of its upper western rim by the hoofs of cattle grazing there. This has long led local lore to proclaim that this hill is a tumulus with an underground passageway under its NW

Figure 2.2. Known locally as Clery's Hill, this mound is said to be an ancient burial mound. It was known as Loovroo in the past.

frontage facing Rath Feerwore. Some elderly locals claim this hilltop mound had been known as Ferach Mor hill, others as Lovroo. One of the latter, an elderly lady with a prodigious memory of now-lost local placenames, Mrs. Smith of Ballyknock, recalled old people long ago referring to it as Loovroo (= Lugh Bhru, Otherworld Abode of the god Lugh). Its position in relation to the former stone encirclement, its close proximity to the original site of the Turoe Stone, to Rath Feerwore and the burial mounds, calls for thorough archaeological examination.

Stone circle surrounding Rath Feerwore

Local tradition is that the entire site surrounding Rath Feerwore adjoining 'Clery's Hill', the apex of Turoe, was surrounded by a wide circle of standing stones. Patrick Keane's grandfather pointed out to his sons and grandsons where the stones of the stone circle originally stood on the NW slope of Turoe. Like the illustrious Turoe Stone itself, some of the stones of the circle were of granite, others were of limestone. According to all but lost Turoe folklore, the outer stone at the W tip of the alignment (which was recorded during the 1915 excavation) was one of the stones of the great encirclement. A similar stone formerly stood very close to this as if forming a narrow ceremonial entrance. Just inside this entrance stood the Turoe Stone, approximtely 10 yards in front of the entrance (according to the publications) to the now demolished Rath Feerwore (Rath Fearach Mhor), but 100 yards according to a 1988 study. One conglomerate block of the circle remains in situ on the W slope, as does a granite stone still standing in the little grove of fir-trees near the NW summit of Turoe. As with Newgrange where only 12 of a total of 34 of the great standing stones remain, so most of Turoe's stones have been carried off. The regally decorated Turoe Stone itself was drawn away, pulled by a team of horses and set up in the front lawn of Turoe House at the E side of Turoe long before the excavations began. Patrick Lyons, who for 40 years had been a herd with Mr. Dolphin, Turoe's Landlord, was on hand in 1915 to point out the original setting of the Turoe Stone.

The graves of Fearta

50 years ago people recalled that the older generation claimed that more than 100 large sepulcral mounds stood the townland of Fearta before the Famine, with less than 50 in 1900. After the Famine many mounds were dug out by starving people seeking valuable grave goods to pay for their passage to America. One such mound was excavated in 1903 by George Coffey. He found an urn with a cremated body, the bones of a horse, a human and some deer bones. From the way they were buried he assumed that an important person was cremated and the human body was buried as a protector of the grave. The deer and horse were worldly goods belonging to the person cremated.

As a boy in the 1930's Patrick Keane of Fearta remembers burial mounds being levelled behind his home in Fearta. He was intrigued by numerous round pits, which he witnessed as a boy in a flat area near the summit of Turoe. These were ploughed out of existence, as was Fearta's last saucer-barrow mound on the N slopes of Turoe soon after the excavations of Rath Fearach Mhor in the 1940's. It was referred to in the report of the 1915 excavations. Another tumulus on the upper rim of the hill between Fearta and Cruach Chon Coolaind was known as Cruachan Airt. Local tradition claimed that it was the burial mound of King Art, son of Conn, father of Cormac Mac Art. This, the last of the burial mounds of Fearta, was bulldozed out of existence in 1994.

Some of the tombs of Turoe and burial mounds of Fearta appear to belong to the Iron Age and be contemporary with the Turoe Stone as suggested by the two which bore the names of Iron-age kings, Art and Conari. A high mound on the E side of Turoe hill is known as Cruachan Shile, dedicated to the goddess Shile. Local children were warned to steer clear of it for fear that the ghost of Shile would swallow them alive. Shile was a Celtic fertility goddess.

Treasures found

The custom of plundering Turoe's tombs continued into the 20th century as old rumours of buried treasure hovered in the area. One find remembered locally was a minature replica of the Turoe Stone complete with curvilinear La Tene art work. It was sold to a 'Yank'. Other finds were a stone head of a three-faced god (a tricephalus) and the skeleton of a huge dog with bronze collar and lead. This final find was made in the 1930's by a returned exile, Martin Scarry, and Joe Ruane. They dug out a massive round barrow mound known as Cruachu Choolan (Cruach Chon Choolaind), at Coolan Hill near the SE corner of Fearta touching the NE boundary of Turoe. They unearthed the skeleton of a huge dog with bronze muzzle, collar and chain. The bronze set was sold to a foreigner. Local lore claims there was an ancient black-smith's forge on Coolan Hill, reminding one of the story of the hound of Coolan, the smithy, slain by Cuchulann.

Ancient roads

It has been said that it was the Celts who were the great road-builders of northern Europe. The ancient roads of Britain, for example, often thought of as Roman, had been laid by the Celts long before the coming of the Romans. This is a fact only now slowly being accepted by scholars in the light of new archaeological finds. It is obvious, looking at Caesar's account of his Gallic campaigns, that he was moving his legions rapidly through Gaul because there was an excellent system of (Celtic) roadways in existence. Similarly, when Caesar crossed to Britain, it becomes obvious to the careful historian that there had to be a well-laid system of roads in existence. Several Roman (Latin) words relating to road transport were borrowed from Celtic into Latin such as carruca and carrus from which also derives 'car' in Anglo-related languages. Also the English word 'road' is borrowed from the ancient Celtic 'rot' as in 'Rot na Ri'

The 5 great roads of ancient Ireland are said to have radiated out from Tara. These were 'Sli Dala', 'Sli Midh Luachra', 'Sli Coolan', Sli nAssail and the Escir Riada. But there is little evidence of ancient roads around Tara. In contrast, there are several existing remains and considerable oral history evidence of ancient roads around Turoe. These include the following:

Sli Dala: Local elders like Joe Cooney, Michael Gleeson, Joe Dunne, Deely, Ruane, and Casserly, were able to point out the route of the Sli Dala for long distances. From the SW corner of Turoe, Slighe Dala continued straight on to Loughrea, leaving sections of pristine ancient highway with banks intact along this route. There

Figure 2.3. A section of ancient road near Ballykeeran Bridge (Factory Bridge). Although narrowed by the construction of a stone wall, the original alignment can still be seen. This road continued south through Ballykeeran townland and on to Loughrea and is known as the Sli Dala.

it swung E on the heights above Loughrea lake to the massive ringforts of Masonbrook where it swung ESE past the Hill of Doon on its way to Portumna, Borris-in-Ossory, Dind Righ near Leighlin Bridge on the Barrow and Roscrea. Stretches of this ancient road are still traceable confirming Joe Cooney and Patrick Keane's alignment. This ancient highway from Knocknadala to Roscrea is still known by many old people as Sli Dala, but it had several local names such as Bealach Mor Muigh Dala further afield near Portumna and E of the Shannon. Other names for Sli Dala on local lips are Sean Bealach, Bealach Mor and Slighe Mor Dala. It was 21 ft. wide and had beautiful broad rounded banks on either side most of which were levelled in the laying of present-day roadways such as those along the W boundary of Turoe, or by farmers through whose lands it passed. A short section can still be traced near Factory Bridge, just north of Turoe (Figure 2.3). Other sections with banks intact can still be seen S of Turoe passing alongside Tonnawausa fort in Lackafinna (Figure 2.4), and again in the field across the road from the Rugby field in Loughrea where one of the best intact stretches has survived (Figure 2.5). Another long section suvives between Loughrea and Masonbrook. It ran in a straight line for miles on end, so unlike present-day roads in the area. The stretch from Knocknadala via Turoe to Loughrea was still used as a shortcut to Loughrea until early in the 20th century when a bridge across a stream was washed away.

Intriguingly, legendary history claims that Slighe Dala ran from Temhair na Ri (Turoe of the Kings) to Roscrea and Borris-in-Ossory. The same history claims that this renowned highway was founded by Dela, alleged father of the leaders of the Fir Belg invasion, who set up his royal seat on the Hill of Knocknadala. He and his Cnoc na Dala, on which Slighe Dala converged, gave it its name. The same legend further relates that Roscrea, its original S destination, got its name from Crea, the Munster Fir Belg Princess who became Dela's consort. Rare indeed the Romeo who so literally cut a royal road straight to the heart of his lady! Dela did. Slighe Dala proudly proclaims so to this day!

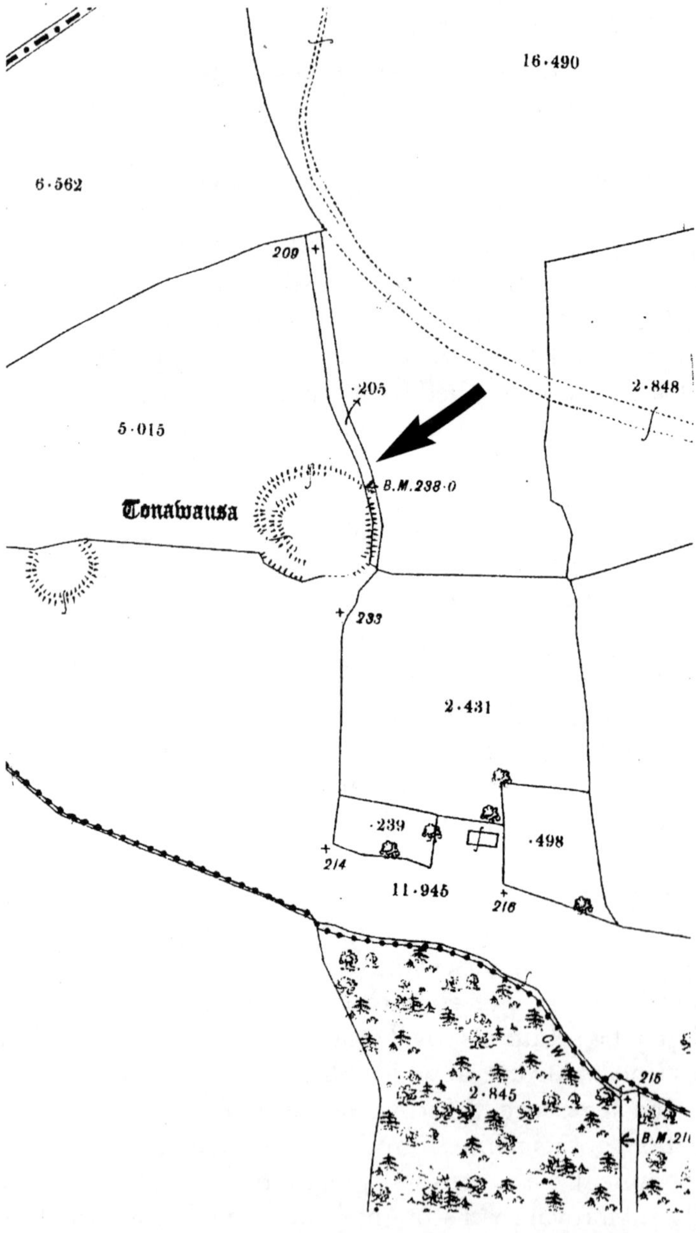

Figure 2.4 Continuation of the Sli Dala at Tonawasa, as shown on the 1894 O.S. map. Map courtesy of Trinity College Dublin.

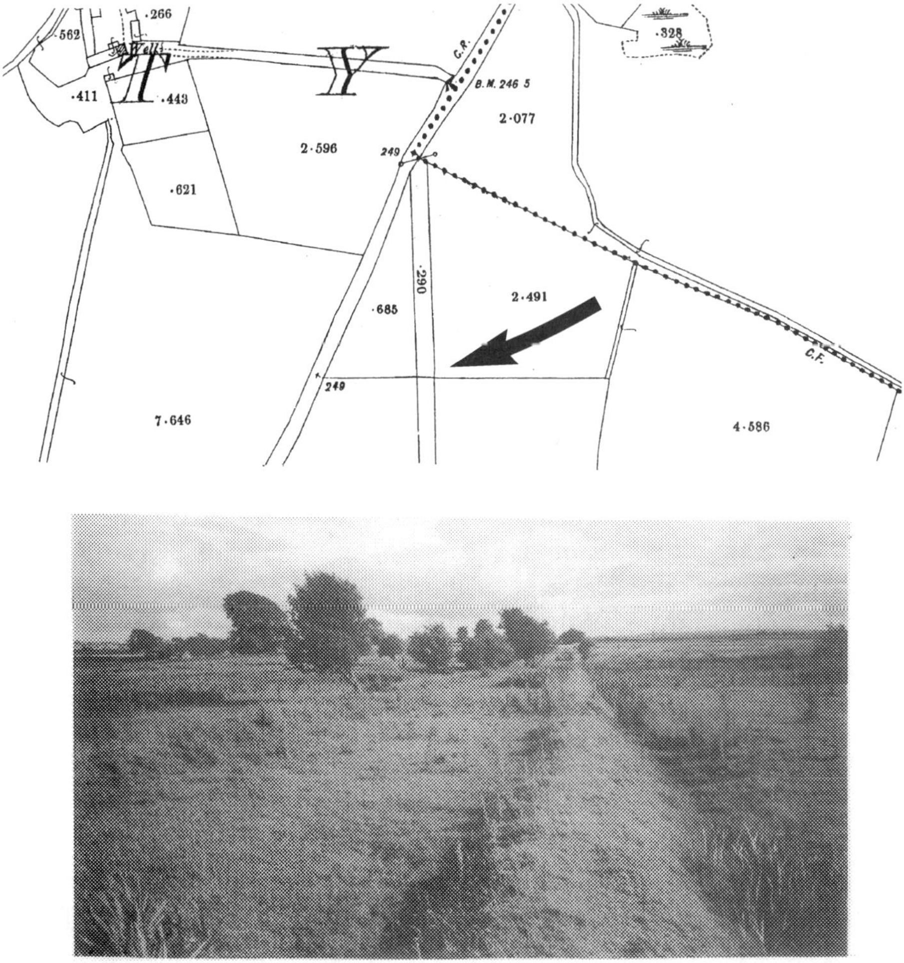

Figure 2.5 Continuation of the Sli Dala at Ballygasty, near Loughrea, as shown on the 1894 O.S. map. Photo insert shows the road as it is now. Map courtesy of Trinity College Dublin.

Escir Riada: Well-informed elders logged the main route of the Escir Riada from the Iron Age port of Ath Cliath Magh Ri (near Clarenbridge) in Galway Bay to the massive Caherdrine fort and thence straight to Athenry. There a branch, Boher Mor, ran W to the present site of Galway city and beyond along the coast into Connemara. To this day an area in Galway is called Bohermore. The main Escir Riada road ran E for 5 miles, through Esker and straight to Knockatogher (hill of the highway; tracable on O.S. maps from Carnakelly to Raford;). There it swung abruptly S through Raford (Rath na bhFille Droichead = bridge of Poet's Palace) and straight on to Knocknadala where it met the Slighe Dala. It then swung east to Shannon Bridge and Dublin along the Escir hills.

A short section, locally called the Escir Riada, still exists in the fields in front of Esker Monastery (Figure 2.6). Part of this was recently destroyed. A further section through Knockatogher is shown on the 1933 O.S. map (Figure 2.7) and remnants of a sunken roadway exist on the North of Knocknadala Hill (Figure 2.8). On reaching the summit of the hill the road continued east, where the continuation along the esker hills was known as the Escir Riada. This road, continuing to Ballefa and New Inn, is clearly shown on the 1933 O.S. map (Figure 2.9).

Queen Medb (Queen of Connaught) had a royal crannogue residence on Loughrea's sacred lake, not on Loch Ree on the Shannon. She made her future consort, Ailill, guardian of this lake. While bathing in this lake she was finally assassinated at the instigation of Conor Mac Nessa. The Escir Riada from Rath Cruacha of Athenry (see later) to Knocknadala and Sli Dala from there to Loughrea were as familiar to Medb as modern roads are to people today. These ancient roads are monuments which must be preserved at any cost.

Rot na Ri: From the point where the Escir Riada met the Sli Dala at the summit of Knocknadala another ancient royal roadway ran WSW (Figure 2.10). Although that section of it which descended W

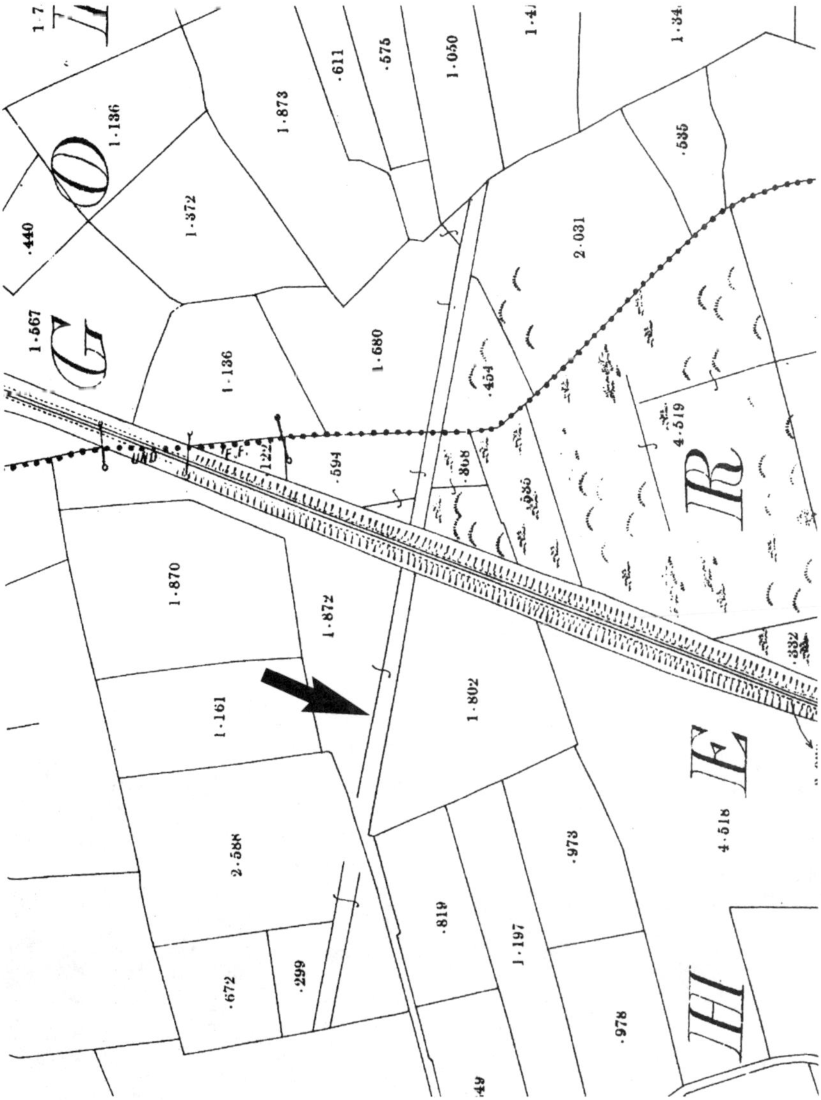

Figure 2.6 A section of a 700 metre section of the Escir Riada at Knockatogher/Gortakeeran, as shown on the 1894 O.S. map. Map courtesy of Trinity College Dublin.

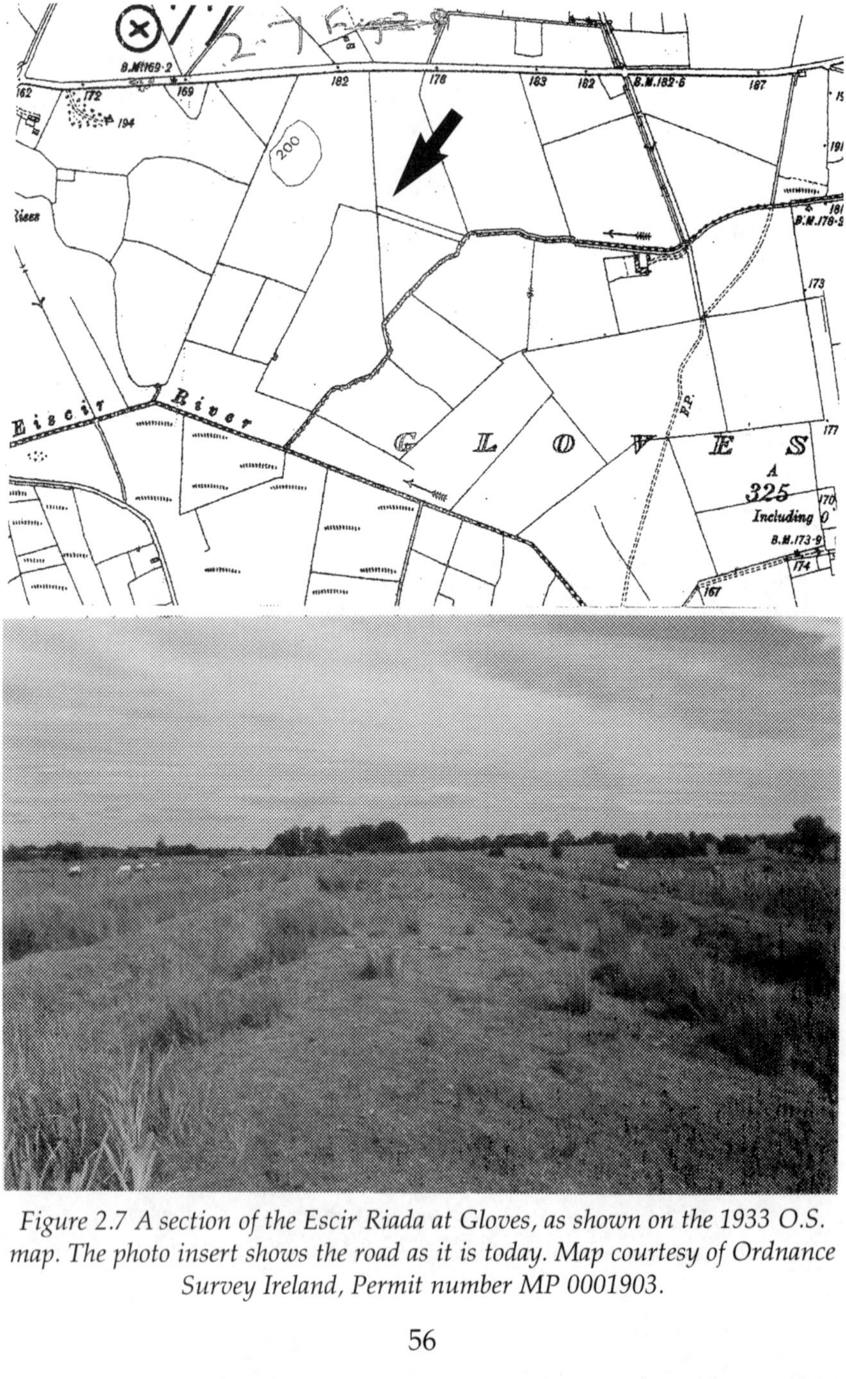

Figure 2.7 A section of the Escir Riada at Gloves, as shown on the 1933 O.S. map. The photo insert shows the road as it is today. Map courtesy of Ordnance Survey Ireland, Permit number MP 0001903.

Figure 2.8. A sunken roadway on the north slopes of Knocknadala Hill. Known as the Escir Riada sections have visibly been filled during land reclamation.

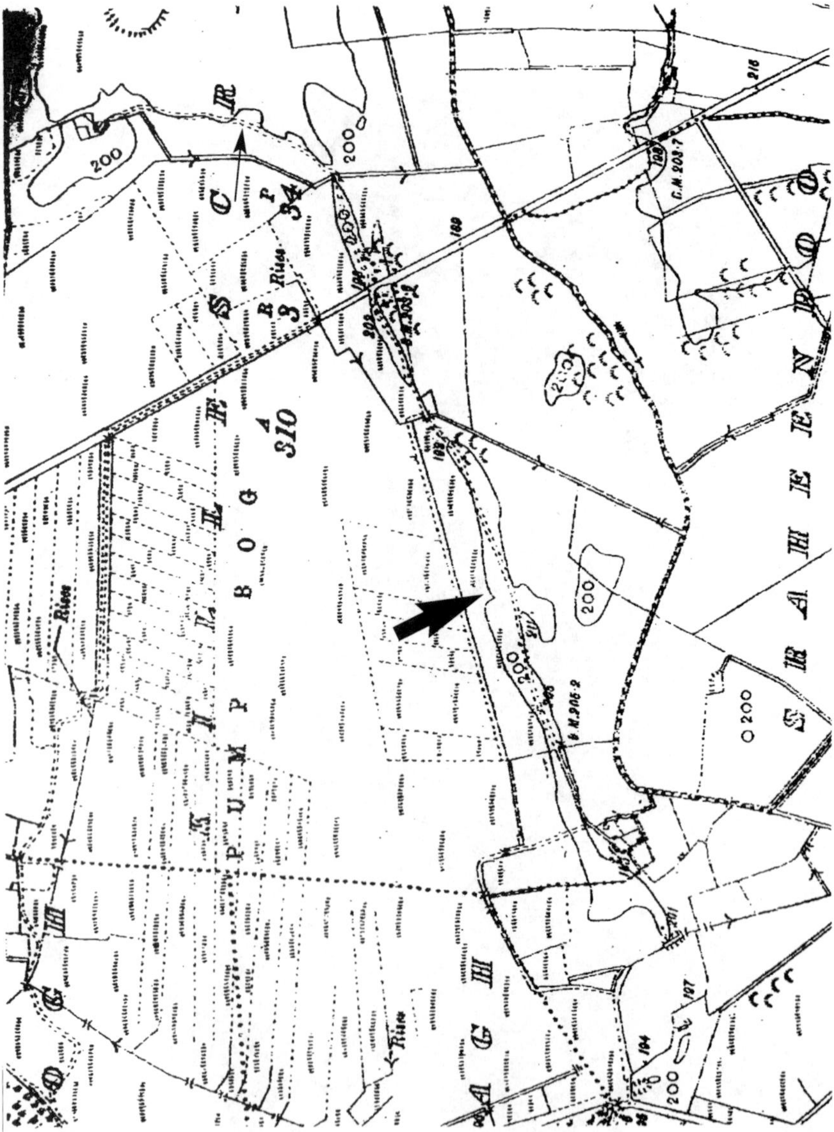

Figure 2.9. The Escir Riada, shown as a footpath on the 1933 O.S. map, going east from Knocknadala Hill towards New Inn. (Map courtesy of Ordnance Survey Ireland, Permit number MP 0001903).

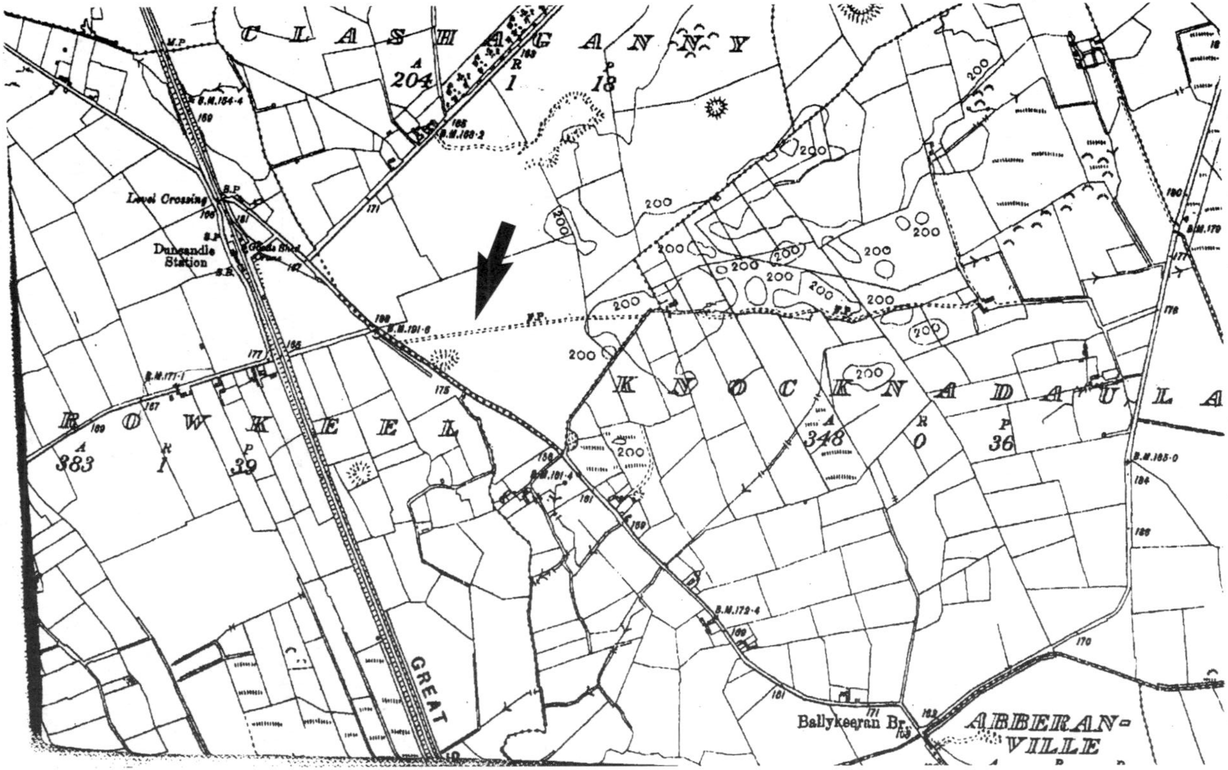

Figure 2.10. Although shown on the 1933 O.S. map as a footpath, this road, called Rot na Ri, is remembered as a banked roadway. (Map courtesy of Ordnance Survey Ireland, Permit number MP 0001903).

from the heights of Knocknadala was levelled out before the 1933 O.S. map was published, its course is still shown as a 'pathway' on this map, although it is remembered locally as a banked roadway. There are apparently two routes from Turoe/Knocknadala to Myode. One came through Cotni (Carrowkeel) and followed the now destroyed esker running W to the river near the Castle in Cloghereavaun and ran W along the N bank of river from Clogharevaun Bridge. It can be followed for a good distance W of the bridge leading towards the very outstanding Esker a bit further west which runs all the way through Greyford, Curraghmore and Rathgorgin to Moyode. The second route starts from Gortakeeran, ran mostly along the present Raford road to the forge at Kiltullagh, along the old narrow road behind Kiltulla Church and on along the road to Kiltulla cemetery. It continued straight on across the fields before joining the main Kiltulla/Athenry road at or near Curraghbeg and then continuing straight on through Rathgorgin to Moyode. From Moyode this road passed through Ballywinna, Parkroe and on to Clarinbridge. This was a shorter route to Ath Clee Magh Ri (near Clarinbridge) than the Escir Riada and may have only been used by Kings, hence the name 'Rot na Ri (Road of the Kings). Many modern byroads lie along several sections of this ancient road and are notable for their straight alignment.

The route of 'Rot na Ri' through Moyode is corroborated in archaic legendary poem called "Magh Fot ar Rot na Ri".

Sli Coolan: This ancient road through Fearta zigzagged as it exited through the inner set of Turoe ramparts and swung into what is today known as the Taillye road in Galboley. For a time it was known as Sarsfield's Road from the passage of Sarsfield's army retreating to Limerick from the Battle of Aughrim but its original name was 'Sli Coolan' according to Deely from Taillye and Joe Cooney of Lackafinna. From the E side of Turoe Hill (in Galbooley townland) the branch of the Escir Riada which converged on Turoe joined the Sli Coolan. This branch of the Escir Riada on which the modern road was built formed the Brackloon Beg/Ben More boundary and the Caraunmore/Cross boundary for a short stretch. There it joined the Escir Riada branch, which climbed over Knocknadala and swung east along its S slope and thence along the now demolished Esker running to Caraunmore and thence on through Ballyfa to New Inn. The royal avenue of Sli Coolan itself swung S around the SE and E flank of Turoe Hill, once again flanked by similar broad banks which characterise other ancient roads. It swung round W through the N corner of Bullaun and across the centre of the S half of Turoe until it reached the ancient Sli Dala running NS along the W side of Turoe. Thus it encircled the Hill of Turoe.

Sli Luachra. Another ancient route branched off from Sli Dala at Loughrea and ran S through Clare. Joe Cooney of Lacafinna and Joe Dunne of Loughrea recall Irish speakers in their young days calling it 'Sli Luachra'. Its Munster terminus was the Fir Belg Capital, Temhar Luachra, 5 miles south of Limerick.

Ceremonial avenue to the sanctuary

From the road along the W side of Turoe is an ancient road marked by a line of granite standing stones (only 3 remain in situ) that ran E through the centre of Fearta on the N slope of Turoe hill. A sunken ceremonial avenue ran S from this road straight up the N slope to the summit of Turoe between Loovroo (Clery's Hill) and Rath Feerwore, original sanctuary site of the Turoe Stone. It has been ploughed out of existence.

The Turoe Stone: The Stone of Destiny, perhaps

Mac Liac's text tells that the Fir Belg Leader, Dela, established the royal Oenach (Festival, Feis, inaugural re-enactment Fair) on a site adjoining his palace on Knocknadala: "Roghni Oenach Dind re thaibh." Adjoining Knocknadala is Turoe hill with its illustrious Turoe Stone originally sited in front of Rath Feerwore above the Cemetery of Fearta. According to 'Dindshenchas' (History of famous places) the Lia Fail (Stone of Destiny) was set up on Cnoc Temhro or 'i Temhra'. At this time there were no High Kings of all Ireland, and early Kings such as Conaire Mor, Conn and Cormac Mac Art, were overkings of the Erann/Fir Belg peoples in the West of Ireland. They were inaugurated at the Lia Fail, which supposedly roared its acceptance of them. As Turoe and Tara have the same Old Irish name, 'Temhair' (nom. case; gen. Cnoc Temhro: dat. 'i Temra') it is easy to see how it is Turoe that is being referred to as the seat of these Kings and not Tara. Indeed, there are contradictions regarding the alleged names on the Tara monuments. If one examines the maps of Tara there are several changes between that of 1919 and the one of 1931. In addition, what is supposed to be the Lia Fail on Tara is actually an upright granite stone marked with a cross, the letters R.I.P. and initials - it marks the grave of some of those killed in the battle of Tara in 1798.

Ireland's oldest origin-legends implicitly indicate that the Lia Fail is of Fir Belg origin. As is now clear and accepted by most historians, until 637 Tara belonged to the Cruthin of Ulster, the aboriginal pre-Celtic people whose art style differed entirely from the highly styalized La Tene Celtic art style of the Turoe Stone. La Tene art is common only in Belgic areas of Britain, Ireland and the Continent, and dates back to the Celtic Kultur-Provinz E of the Rhine, the cradle of the Belgae.

The Turoe Stone is covered with continuous abstract curvilinear La Tene art with flowing design carved in relief, set off by a band of Greek-key decoration surrounding the foot of the stone. Resembling the famous 'Naval Stone' of Delphi with its Greek-key design, it reminds one of the ancestral branch of the Fir Belg, the Volcae Tectosages, who went down into Greece and desecrated its sacred shrine at Delphi. On Turoe hill stands Celtic Europe's finest example of La Tene relief carving on stone. From the conspicuous place given to it in early records, it is evident that the Lia Fail was unique. The Turoe Stone certainly lives up to this uniqueness.

Rath na bhFhille: Poets Palace; Aber na bhFhille: Poets Arena

Every Royal Assembly site had a place set aside especially for the Poets, Bards and Literati. Poets, bards and minstrels held high position in Celtic society and were closely associated with the druids who included many tribal chieftains. Poets and bards were a highly trained professional group who were the repositories of Celtic history, legend, folklore and poetry. They were under the patronage of the chieftains. The tradition was strictly an oral one, the bards having to commit to memory a vast store of knowledge and be word-perfect in their recitations. Greek and Roman writers lauded the refinement and elegance of their use of language and appreciation of linguistic subtlety. Poets played a privileged part in the great assemblies where they had their own arena and apartments. It was designated Rath na bhFhile, the Palace of the Poets. The Turoe/Knocknadala complex was no exception. There are in fact two sites, one on the NE slopes of Knocknadala known as Rath na bhFhille Drochaid (the name of Raford townland in early records). The other site is on the NW slopes of Turoe, between Fearta and Knocknadala, called Aber na bhFhille (clearly pronounced as such locally, but written as Aberanville on O.S. maps), Forum, Field or Arena of the Poets, where their literary feats were contested. The close relationship between these townlands is shown in Figure 2.11.

This completes the summary of sites around Turoe. The varied and extensive nature of these sites, the Celtic connotations and Royal implications, in addition to the complex series of embankments surrounding them, emphasise the significance of this area in ancient times.

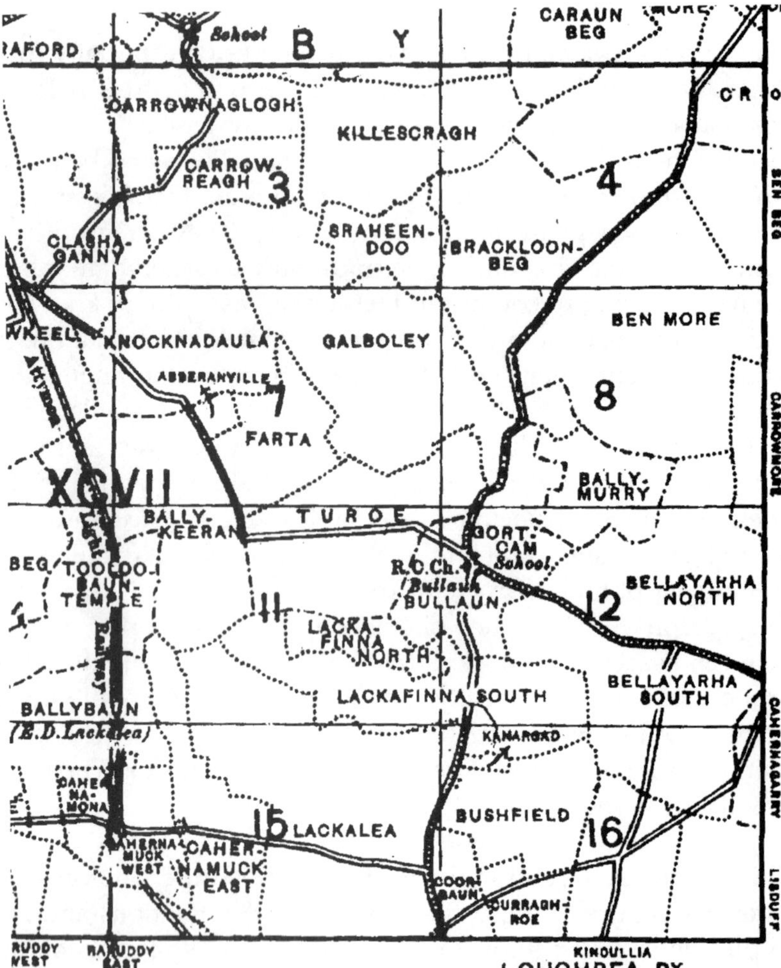

Figure 2.11. Map showing the close relationship of townlands with significance of celtic centre of power.
(Map courtesy of Ordnance Survey Ireland, Permit number MP 0001903).

Chapter 3.

Extensions of the linear embankments around Turoe/Knocknadala

Second extension of the linear embankments around Turoe/ Knocknadala outer ward

As we have seen in Chapter 1, an extensions of the outer ward linear embankments surrounding the Turoe/Knocknadala complex can still be reconstructed. As detailed in this chapter, further extensions can also be reconstructed.

Working once again from the SE corner, 5 miles SE of Turoe, from Carra just S of Kilreekil, the linear embankments ran W straddling the Cooleeny/Ballynahistil and Drought/Ballydoogan boundaries. Through the former Black Wood (Coill Doo = correctly Dyke Wood) they ran astride the Kilmeen/Moyleen boundaries with Ballybroder. The set left its trail W through the S of Kinkullia and the N of Fairfield and Caheronaun with a string of forts guarding the line of ramparts as they raced through Fairfield commanding wide vistas. Ramparts in Graigue were effaced in the development of Loughrea Golf Course. So too was the site of an archaic village complex of stone-slab housing and a narrow-gauge sunken road system with 4 foot wide walls (shown on the O.S. map; partly preserved outside the Golf Course) linked to the Sli Dala in Ballygasty. A bulldozer carrying out the work uncovered an underground bunker along a line of rampart (Figure 3.1). Tusks of the mythological wild boar rooted up the ramparts W through the S segments and bogs of Curraghs, Ballygasty, Raruddy East and West, and the N parts of Monearmore, Cosmona and Gorteennabohogy. The set raced W across the marshlands in the N parts of Caherlavine and Cahernagormuc. It crossed Ardnadoman East/West (Ard na Domnann, Hill of the Domnann, a Celtic tribe

Figure 3.1. The 'cave' entrance near Loughrea Golf Club, which was discovered during excavations.

who defended this oppidum flank). There is another underground bunker in Ardnadoman hill. The set skirted Cahertinny, Gortsheela and Cahernaman as it ran through Carrowmore.

It swung NNW at Ballnamucka along the Knockdaumore (Dyke hill)/Srah boundary, Cloghastookeen and Doogaraun. It ran through St Cleran's, straddling its boundary with Derryhoyle Beg and Coolraugh (outer dyke). Stream banks through these townlands were raised and fitted with floodgates to flood stretches of land behind them to enhance the defences. Departed elders Martin Quinn, Pat Curley and Mr. Martin related old traditions of Tinny and his Fianna fighting men at Cahertinny, Ardnadomann, Caherlavine and Cahernaman, centres of the Dumnann and Manapii of the Colla Mann branch of Moycola in the service of Cormac mac Art. The presence of the Fianna fighting men in the Loughrea/Aughta region reverberate in local legend. The Yellow Bog road through Cloonoo, which was built on a turf based rampart, disintegrated. The laying of the new road to replace it removed all vestiges of the ramparts.

It sped through the E of Lickerrig, along the Lacarrow/ Ballynahivnia boundary to Glebe and its boundary with Lisalondoon, skirting Cahercormac and Garrakyle. Cahercormac is said to have been Cormac Mac Art's fort when he retired from the Kingship after being blinded in one eye and after his rejection by the Druids for having become a Christian. It lies within the E edge of the ancient forest of Achaill Oichne, by which this area was known in ancient times. This is a fascinating example of how the oral history of this area is corroborated in history. In pseudo-history, Cormac is said to have retired to Achall, which has always been claimed to be Arraklye near Tara. But yet this area of Galway is rich in oral history relating to Kings who were supposed to live at Tara, but of whom little evidence can be found there. If Cormac had really been King at Turoe, it is easy to see how he would have retired to Achaill Oichne a few miles away.

Figure 3.2. Avenue-like tree-covered rampart in Rathgorgin townland.

Lisalondoon, another townland in this area, may retain the name of the city of London, capital of a Celtic Tribe called the Trinovantes. London was originally called *Lug Dun*, but its name was changed to *Lud Dun* (degraded to London) after Lud who rebuilt it after the Catuvellauni, another Celtic Tribe, captured it from the Trinovantes. When the Trinovantes fled Britain with the Iceni and the Cantii they may have sailed to Ireland. Townlands with elements of the names of the Cantii, Iceni and Dumnoni occur widely in the area adjoining Lisalondoon. A forested area on the E edge of Achaill Oichne forest was known as *Coill Clann Iceni* (Iceni Forest, corrupt Kilconicny). An area ajoining the same forest stretching E to Lisalondoon is still known today as Treanchoill, which could be interpreted as 'The Trinovanten Forest'.

The ramparts sped on through Hollypark, Cregg, Kilconierin and the W side of Carrownamorrissy along the Dooeyertha (Dughiortach/*Dur*) River into Rathgorgin (Figure 3.2) where they marched with an outer ward rampart of the Athenry oppidum (see later). In William Larkin's 1819 Map of Co Galway, which has a number of placenames not on the O.S. Maps, there is Cahernamorrisy in the place of Caherakileen/Oldcastle. Among other interesting names on this map are Thollaghree in the place of Kingsland North in Athenry and Shanghla (Sean Chladh) near Derrydonnellbeg Cross.

Carrownamorrissy is one of a few similar placenames in the area, such as Lisnamorrissy, along this W front of the Turoe/ Knocknadala oppidum, named after the Belgic Marrasaii. In this area the townland names and placenames again corroborate legendary history. Lughaid Mac Con came from Scotland with a large fighting force. Among these were the Marrasaii, Menapii and Fotads (Votadini). These tribes have left their names in the form of townlands such as Carrownamorrissy (Marrasaii), Ballymanagh (Menapii) and Moyode (Magh Fotad = Plain of the Votadini). Corc (also called Gorgin) was one of their leaders and thus there is Rathgorgin today. One of the large forts in the area is called Clochar na ngFhotad and another Lis na Gal (= fort of the foreigners). Lughaid Mac Con was driven south into Clare and across the Shannon into Munster. Legendary history relates how Cormac mac Art suffered an embarrassing defeat when he went south to aid his Munster allies who were being overrun by Lughaid Mac Con and his fighting men. Corc had stayed behind at Rathgorgin, but eventually went on set up the Munster royal seat of Cashel. Corc's palace in Rath Gorgin, surrounded by the old river channel, was converted by the Normans into a motte and bailey castle, which still survives intact.

Across the Dughiortach River this outer ward extension met the set of ramparts of the Athenry oppidum which ran NNE from Rathgorgin through Esker, Killascaul, Gloves South and Gloves Middle to Clonbrusk. From there the Turoe/Knocknadala set ran ENE astride the Gloves East/Clonbrusk boundary. Carnakelly bog road zig-zagged between the inner ramparts. The outer rampart formed the Cloonkeenmore South/Carnakelly North boundary. The name of the string of 'Gloves' townlands is *'Na gChlaidhbhe'* (plural of 'Cladh' = defensive dyke), the dykes being well known to the older generation. This name proclaims the former presence of the dykes where they are demolished. The many Gloves townlands were the meeting place of the Athenry outer-ward dykes and those of the Turoe/Knocknadala oppidum, overlooked by Dominick's Hill, two coliding oppida with different alignments each accommodating the other.

The phenomenon of bringing bog, forest and other natural defensive features within the fortifications of the oppidum defence was a military strategy, which will be encountered again and again in later extensions. Parallel transverse ramparts survive along the E edge of the bog along the Laragh road running N to Laragh Cross Roads. One was known as the 'Carriage Way' according to the late Ned Burke who lived across the road from it in Laragh from where it ran NNE to the outer rampart on Laragh Hill. The narrow Lis duff/Attymonmore neck of land was a strategic passage-way into the oppidum. There was no other entrance from the N across the expanse of boglands for miles on either side.

Hence it was heavily defended by several ramparts instead of the more usual three.

The ramparts (Figure 3.3) ran across Attimonmore South (*Ait Ui Maine Mor* = Place of Maine Mor's people). They ran astride the Cappanashruhan/Killimor boundary, through Cappananool, Clooncah and Streamsford with its surviving massive rampart forming its N boundary, skirting its turret forts. They swung SSE through Skehanagh, Coppanagh and Gortmore, straddling their boundaries with Beefield and Gortlemon. They sped through Tooreen and Castlebin North, South and East, skirting Castlebin Fort and several others along the E side of New Inn's Knockbrack and Knockmore. They pressed on through Treanbaun, Highpark and the E protrusion of Ballyglass into Glennaskehy bog where they swung S by SSW through Ballintubber West and East into Newgrove. Impressive vestiges of linear and transverse ramparts shown on the O.S. 6" map Sheet No 98 survive in Rathglass, Carra Bog, Balintobber and Newgrove. In the latter townland one low spoilt vestige is "wide enough for two lorries to drive side by side on top of it," as local farmers describe its width. In land-reclaimation throughout the entire oppidum area, excellent vestiges are being wiped off the face of the Celtic landscape by modern machinery.

Figure 3.3. Section of ramparts running across the bog near Attymonmore South.

The ramparts ran astride the Doon/Carnaun boundary, leaving vestiges along the Dublin/Galway road. Vestiges of transverse ramparts survive along the Bullaun/Kilreekil road forming the Doon/Wallscourt boundary. The turret-like fort in Lecarranagappogue in Kilreekill (Chill Rechil = Church of St. Rechil) served as a look-out post on the line of ramparts. The inner rampart skirted Toberahoney (Tober a Shonnaigh = Dyke well) and a nearby high fort in the NE of Wallscourt. The rampart set straddled the Wallscourt/Meanus, Dartfield/Glenmeen boundaries and ran into the W side of Carra where they swung west to the starting point.

Third extension of the Turoe/Knocknadala outer ward

While the Turoe/Knocknadala oppidum was expanding apace a new development arose on its NW flank. A new oppidum emerged, centered on Athenry town. Its inner ward linear embankment survives as the well-known earthwork along the N and E walls of the town. Numerous vestiges of the linear embankments of its outer wards survive through a 5 mile radius of countryside. The outer ward ramparts of the Athenry oppidum, known as Rath Cruacha and Magh Muc Dhruim in archaic Dindsenchas tracts, and those of the Turoe/Knocknadala oppidum were destined to collide as they expanded. This point has already been reached in the reconstruction of the outer wards of the Turoe/Knocknadala oppidum and will affect, and be reflected by, the further reconstruction of the oppidum. To cater for this, the two expanding oppida accommodated each other. The E/W Dooeyertha (*Dughiortach*)/Craughwell/ Dunkellin River and the NNW rampart alignment from Greyford in Kiltulla (SE of Esker) along the W boundaries of Carnakelly and Cloonkeen to the Killaclogher River became the boundary line between the two, reflecting a satisfactory solution when the two oppida collided. Later extensions of the Turoe/Knocknadala oppidum enveloped the Athenry oppidum.

Reconstruction of the third extension of the Turoe/Knocknadala oppidum begins this time from the NW corner and proceeds clockwise (See Figure 3.4 for an example of the ramparts here). Starting 4 miles ENE of Athenry from the N tip of the Cloonkeen/Ballyboggan boundary straddling the Cloonkeen river where it marched with an outer-ward extension of the Athenry oppidum, the set of ramparts ran E, slightly ENE. Its alignment was dictated by the narrow strip of land between the bogs embracing Cloonkeen through Cloonkeenmore North. The name Cloonkeen on several townlands is a corruption of Clann Chian, descendants of Cian. He was Fir Belg leader of vassel tribes of fighting men (Collaibh Oige, corrupt 'Colla Woge'), who came from Eli in north Munster seeking military service with Queen Medb in the expansion of the Turoe/Knocknadala and Athenry oppida. Their task was to erect and defend the fortifications of the outer wards along their north flank. They were rewarded with grants of land given to them for their services. They formed an extensive buffer state between the Fir Belg oppidum and the Sogain branch of the Cruthin. The Cruthin occupied the country stretching all the way from this outer ward of the Fir Belg oppidum at the centre of Co Galway right up across NE Connaught and the whole Northern half of Ireland. With each extension of the outer ward, the Cruthin territory contracted rapidly. After Maine Mor's 5th century invasion of their lands (see later), the Clann Chian took up service with Cormac mac Art against the Cruthin of Ulster. They were awarded prime lands in payment for their military services where they later became known as the Chianachta and survived there well into the middle-ages along the N Leinster coast. Although formerly far more extensive, Cian's Connaught territory was drastically reduced to 14 Clann Chian townlands after Maine Mor's invasion.

From Cloonkeen village on the N bank of the Cloonkeen River the set of ramparts cut their way E through Cloonkeenmore North, Ballynanulty, Attimonbeg, Gortnalone South, Cappanasruhan and Clooncah where the Dublin/Galway railway line swung more ENE to take advantage of one of the ramparts. The names

Figure 3.4. Section of ramparts running from the third extension of the Turoe/Knocknadala outer ward.

Ballynanulty and Attimon are of 5th century origin. They make a profound statement of political significance to Connaught history and, consequently, to the wider Irish context, which shall be explained later. Suffice it to say here that Attymon (Ait Ui Maine) is named after Maine Mor's people. He led a massive army from Ulster into the lands of the Fir Belg in the late 5th century. According to the invasion story preserved in the life of St. Grellan, Maine Mor's army marched towards the centre of the Fir Belg stronghold. The route they took was along the barricaded passage through the boglands leading in from Gurteen to Knockaboy (Cnoc an Buaidh = Victory hill), site of one of the placenames mentioned, Knockasheefraigh (Cnoc an Shee Fraoich). The Fir Belg tribes of this part of Connaught, who were under the control of Cian, were annialated, supposedly swallowed up by the bogs, as the story says. After the victory, Maine Mor's followers settled in the Attymon area and in the nearby townland of Ballnanulty (Baile na nUltaigh = abode of the Ulstermen). Ned Burke of Tample claimed the original name of nearby Gortnalon was Gort na Lann (field of the spears) from the number of spears found there after the massacre. The 'Attymon Hoard' (Figure 3.5) of bronze La Tene horsebits and pendants in the National Museum was found at the edge of Cappanashruhan bog. Pointed stakes planted in the bog to impale cavalry horses were found in turf-cutting operations. Mass burials have been unearthed on gravel hills near the scene of the massacre of the Fir Belg by Maine Mor. In the shadow of Knockaboy lies Air Cealtragh (Caltra massacre), Clooncah (battlefield) and Moyarwood (*Maigh air* = plain of the massacre), all announcing the killing fields of Maine's massacre of the Fir Belg.

The rampart set raced from Clooncah to Dooghcloon (Dyke field; 'doo' in Old Irish = dyke) and Derrydoo (Dyke oakwood) Lough, through the S tip of Moyarwood and Greenhills and ran astride the Carrowholla/Carrowmore boundary with Woodlawn to the road and railway station. It swung SSE astride the Woodlawn road, partly built on a levelled rampart which formed the Cloonahinch/Kilaan boundary with Woodlawn. It cut through

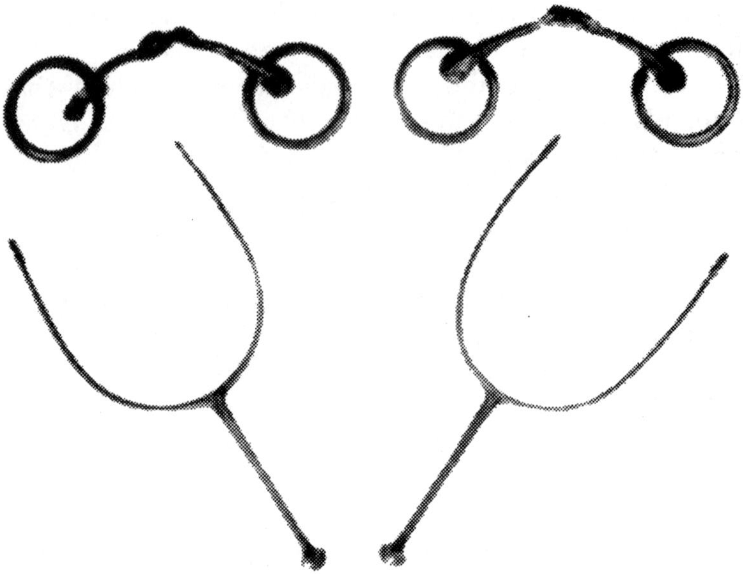

Figure 3.5. The Attymon Hoard, a decorative bronze La Tene horse bit found in the bog near Attymon.

Cloonymorris. It sped through Lisnamoltaun and Killagh More where vestiges survive. One was known as Escirgclaidh (sand-hill dyke) before it was mined for its sand. The set ran through Ballymabilla (ancient forest) skirting Ashbrooke, Turksland, Toormore and Doonaree as it sped to Eastwell. Some townland names proclaim that it ran through a heavily forested periphery surrounding the Turoe/Knocknadala oppidum, corroborating Caesar's description of the classic Belgic oppidum.

Doonaree (Kings' fort), Cartron Sheela (Abode of the fertility goddess, Sheela, associated with royal sites) and Tooree (King's hill) signal a minor royal centre here. The rampart set straddled the Ballydonnellan W boundary, the E of Cloonmain and Lissalumma and the Carrowreagh/Corbally Mor/Lisheenahevnia boundary with Ballyhogan. It sped over Finnure (Cnoc Findabhair = Findabhair's hill, named after the youngest daughter of Queen Medbh and consort of Fraech mac Fidaig, King of Connaught). Like the magic boar the rampart set tore its way through

Castlenancy to Ballynamurdoon (*'Mur Doon'* was a Cruthin term referring to the dyke defence line corresponding to the Fir Belg *Dun Cladh*). The Dal nDhruitne (ngChruthine) of this area were a Cruthin clan descended from Celtchar mac Uithechair. They were isolated from other Cruthin tribes in the Fir Belg invasion, and resided SE of Loughrea.

From Gortymadden the rampart set sped W through Carrowroe (dyke quarter), Ballydavid along the S slopes of Ballydoogan (Dyke townland) hill, Knockbaron and Carrowmore. It straddled the Carn/Knockadikeen boundary (*'dikeen'* is a corrupt Gaelicization of 'dyke' proclaiming its passage, as does the following townland of Tonnaroasty (Tonna rooasty = bottom-land of the dykes). A vestige of a dyke and fosse survives in the valley in Baunoge through which water from the surrounding hills drained into the N end of Loughrea lake. Surface traces of the central rampart survive along the slopes on the Baunoge/Gorteenapheebera boundary. The set straddled the Tonnaroasty/Gorteenapheebera and Farranalynch/Caherwalter boundaries, and skirted the N shore of Loughrea lake. The building of this Norman-walled town demolished traces of the ramparts in that area. The North Wall fosse may be a vestige of the Iron Age Belgic fosse. The set of dykes ran W from the lake through Cuscarrick and Pollroebuck (Poll Roo = Dyke Bog) while the inner rampart ran through Cosmona. The set ran W through St. Lawrencefields and Tullagh Upper and Lower. Like the magic wild boar mythologically associated to the construction of the fortifications, the ramparts cut through Cloonoo (Cloon na nDhoo = Dyke field), the Knockauncoura (Cnoc an Couragh = hill of the entrenchment)/Glenatallan boundary area and the SE tip of Ballnamuca. Sheamus Callanan of Knockauncoura village pointed out vestiges of ramparts running through the Knockauncoura and Cloonoo area and flexed his wide knowledge of similar dykes in several parts of Galway, which he hunted widely as a youth. Such wide-memoried men have saved these ancient linear fortifications and all their historic ramifications from being lost forever.

Figure 3.6. A view over Bahasane Turlough.

The ramparts straddled the Gortawullaun/Lurgan, Boherduff/Tonbaun/Carrowclogh and Cartron/Carrowmunna/Moycola boundaries, skirting Gortnabarnaboy (victory gap). Moycola (plain of the Colla Menn, a Menapian vassel fighting men), is a shadow of its ancient extent, stretching from Loughrea to the Iron Age seaport of Ath Cliath Magh Ri near Clarenbridge. When the sea was 6 to 8 feet higher from the 2nd to the 5th century, it extended inland to the strongforts along the shores of Dur Loch (now the Turlough of Rahasane on the *Dur fluvium* (*An Dughiortach*) of Ptolemy's record) between Crughwell, Clarenbridge and Maree (Figure 3.6). The vassel fighting force manning the lines of the various outer wards of this sector of the Belgic oppidum was made up mainly of Belgic Manapii (Manach) who left their name on many townlands which herald their early presence. Examples of these are Mannin, Manninard, Carrowmanen, Clan Manach, Ard na Manach (wrongly anglicized as Monksfield), Ballymannagh, Newmannin (corrupt anglicization of Mannin Uagh = Manapian necropolis) and Carrowmunna. With the higher seas of the early centuries wide areas S of Rahasane, Craughwell and NW of these around, Foirchoill ('wood of the sailor's rest' according to Irish speakers; wrongly anglicized as Coldwood) and Moyveela (Magh Mell = sea plain) were inundated by sea. As such, the area closely resembled the former lands of the Manapii in the Rhine Delta and those occupied by them along the shores of Loch Erne. These Colla Mann were one of the 3 Collas (Collaibh Oige = young warriors), repaid in the lands of their buffer states for their military services to the Turoe/Knocknadala Kings.

From Ballymoneen and Ballybaun the set swung NNW through Seefin, skirting its Knockroe (Dyke hill) and Toorclogher boundaries. Tradition claims Toorclogher (Temhair Clogher) had a fort near its summit known as Temhair Clogher Nath I (nDhathi), palace of Dathi mac Fiachrach, 5th century king of Connaught and Fir Belg Over king. Pseudo-history claims he was High King of Tara (*Temhair*). His '*Temhair*' was right here in SW Galway, not in Tara of Meath. The Toorclogher/Seefin area was the centre of Ui Fiachrach Aidne, the kingdom of Dathi. Dathi was a descendant of the Fir Belg kings of the Turoe/Knocknadala oppidum. In the adjoining townland of Blackgarden is Toberneighey (Nath I/Dathi's Well). Toorclogher was the site of Diarmad and Grania's Bed. It was arranged for Grania, daughter of King Cormac Mac Art to bethrow an elderly Finn Mac Cool. She eloped with Dermot (Diarmad Mac Morna, brother of Gol Mac Morna, leader of the Connaught Fianna) but was ambushed by Finn in their forest hideout. This became known as Fidh Mac Cool and the adjoining townland was anglicized as Mac Cool's forest or simply as Cool's Grove today (no grove survives).

The rampart set raced through Blackgarden, corrupt translation of Garra doo (Dyke park), Derryhoyle More and Reaskmore. Derryhoyle More (Doire Choill = forest) and Reaskmore (great marsh) proclaim once again that the defences of this Belgic oppidum incorporated both forest and marshlands. It traversed Crossderry, Capanraheen, Creggaun, Loughgill and Coolraugh (rampart end) on the bank of the Craughwell river and in the fork of Dunkellin river. Ramparts across the river belong to the outer ward of Athenry oppidum, with which they collide in the narrow fork along the Dur river (Dughiortach/Dunkellin river). In this area of dense forest, as the townland names proclaim, each townland in the fork of the river makes its own political, geographical, historical or topographical statement. Treankyle (Trean Coill = forest of the Trenovantes) is within the two arms of the Dunkellin river. The massive townland of Ganty (na gCantii) preserves the name of the Belgic Cantii who came with Verica/Ferach, grandson of Commius (Gann), one of the early Fir Belg leaders. They had a massive fort here as their base. It was the original Caherkinmainwee, the name of the small adjoining townland. This name (Caher Chinn Main Mhuigh = fort at the head of the plain of Maine, son of Queen Medb) makes the strongest political, historical, geographical and topographical claim. Maine's residence stood at Clogharevaun near Kiltullagh beside the core of the Turoe/Knocknadala oppidum. He succeeded his mother Medb as monarch of Connaught. Main Mhaigh was then the same size as the Turoe/Knocknadala oppidum enclosed within the set of ramparts of the present outer ward under reconstruction. It is described in Dindsenchas texts as the Plain around Loughrea. Rath Croghan of Roscommon, fictionalized by pseudo-history as the royal capital of Connaught in Medb's time, was still part of Ulster. The profound statement made by this tiny townland, which demarcated the W extent of Magh Main, is very significant. Coill Winard in adjacent Cappagh further corroborates the claim that the fork of the river was a heavily forested area in ancient times. Doonard, overlooking Craughwell, Caherdangan and other strong forts in the area stood sentry along the high ground above the river, defending its crossings. Across the river the ramparts marched with those of the outer ward of Athenry oppidum up to the starting point.

Fourth extension of the Turoe/Knocknadala outer ward

Before reconstructing the Athenry oppidum linear embankments with which the outer wards of the Turoe/Knocknadala oppidum were now colliding, one further outer ward extension of the latter oppidum is reconstructed here.

Working from the Shoodaun/Templemoyle boundary and Newcastle church area, a new set of Turoe/Knocknadala linear embankments ran E skirting the forested bogs astride the Cloonkeen Beg/Cormacuagh boundary. This continued through Clough (Poll a cladh = Rampart bog; the dykes are shown on early O.S. maps; Figure 3.7) parallel to the E extension of the Killaclogher river. From Tample it sprung over Knockaboy, Sheeaun, and Caltragh (*Aircaltra*), skirting Attimany. Each of these townlands and nearby Caltraghbreedy have their own story to tell. Sheeaun was formerly known by locals as the Otherworld Abode of the god of eloquence and learning, Ogma, son of Dagda who was Father and Ruler of the Celtic gods. In Celtic literature Ogma retired to his 'sidhe' on the hill of Airceltra (Caltragh). Tradition claims this was on the summit of Knockaboy, originally known as Sheeaun. Caltragh became *Air Celtragh* (Massacre of Caltragh) from the massacre of the Fir Belg by Maine Mor in this arena

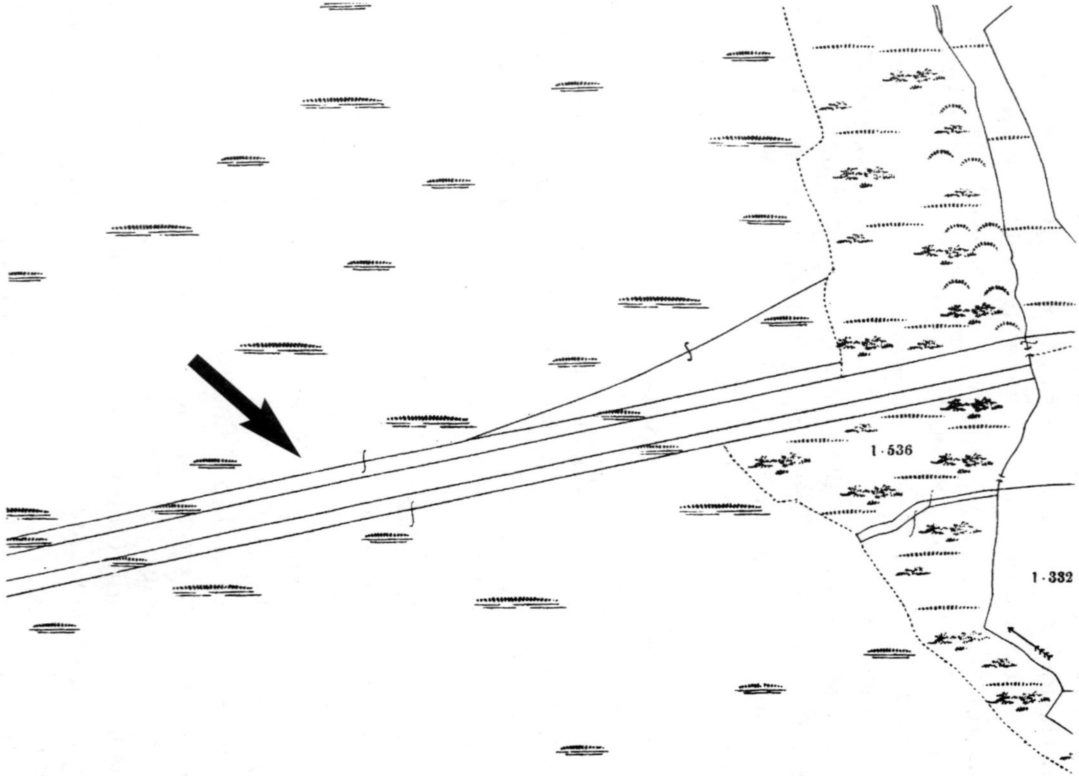

Figure 3.7. A section of rampart running for at least 1 kilometer through Clogh townland. Compare this with the section running through Clonbrusk, which was described by Martin Finnerty in the 1930's. Map courtesy of Trinity College Dublin.

between Clooncah and Caltragh. Ogma's sister, Brigid, goddess of fertility, poetry, healing, and smiths, had her Otherworld Abode in nearby Caltraghbreedy (Caltragh Brigide = sanctuary of the Celtic Briganten goddess Brigid). This sanctuary was then in the lands of the Clann Cian. In this sanctuary of the Celtic deities an early church, Teampail (Tample), was erected. In local tradition it was associated with Cormac Mac Art and claimed as the place of his burial mound (Figure 3.8) in the archaic Tample cemetery (destroyed except for the mound) as attested by the name Cormacuagh (Cormac's tomb), which included much of present-day Clough (Pollacladh) townland.

Figure 3.8. This burial mound near the ancient cemetery at Tample is reputed to be that of Cormac Mac Art. The townland name Cormacuagh (Cormac Uabha; Cormac's tomb) derives from this.

The rampart set sped through Lenareagh and straddled the Clooncah/Caltraghbreedy/Dooghcloon/Moyarwood boundary with Gortronnagh, Fahy, Shanballard and Corskeagh. It crossed the N parts of Greenhills, Carrowholla, Carrowmore and Moneyveen. Skirting the boglands of Liscune the set of ramparts swung SSE through the E side of Moneyveen. Straddling the Clooncalleen/Hillswood, Clooncallis/Ballyglass and Corraneena /Ballintober boundaries it straddled the Killagh Beg/Hazelfort boundaries. It sped on through Ballinphuill and Ballynaclogh, skirting Woodberry, guarded by several forts. It straddled the Sliaun Beg, Gortnahimrissan and Mountain boundaries with Cappataggle and the Pollatlugga/Brackloon boundaries with

Figure 3.9. Avenue-like ramparts provide shelter for animals whose hooves plough up the earth which is then blown by the wind and washed away by the rain.

Rooaun (ramparts) and Tooree. It ran astride the Chelsea, Poppyhill, Ballinrooaun (dyke townland) boundaries with Ballydonnellan East, Lissacullaun, Carrowreagh and stradd- led the Corbally More/Corbally Hogan boundary into Mullagh More (near the School) and Lismacteidg. It straddled the Foxhall Little (Sonnagh Beg) boundary with Cappanaghtan and Foxhall. *Sonnagh* is the Old Irish word for a defensive rampart through bog or marsh, not to be confused with the Irish word for fox = *'sionnach'*, although foxes did hole up in ramparts. Rampart vestiges survive there and in Lurganshanny (Lurgan Shonnagh = long dyke). It crossed Liscoyle, Drumeyre, Drumhogan, Boleyroe, and Coolagh.

It swung W through Killnamullaun, straddling the Rahyconor/Cormick/Lisduff (-doo = dyke fort)/Garaunnameetagh (moote-land), Cloonprask, Gortaneare, Shanvoher, Ballindrimma/Shangarry boundary, Cloghbrack, Clogharoasty and Lissaphuka. It straddled the Rathfarn/Kilmacrah and Srahdoo/Carheendoo (dyke fort) boundaries. Floodgates were set on the banked stream. Skirting Lissheenahasty (dyke), it straddled the Glenaslat/Traskernagh boundary. Through Masonbrook several forts stood sentinel, one of which is Rathsonny (dyke fort). It proceeded through Moanmore, Loughaunlea (rampart lake), into Moanmore East. The stream through Moanmore, forming the S boundary of Moanmore East, had raised banks and sluice-gates to enhance the defences by flooding. The Masonbrook/Loughrea area was a major centre of the Dal gChruthine (Dal nDruithne), a Cruthin enclave related to the Sogain of E Galway and the Cruthin of NE Connaught and Ulster. They claimed descent from the Ulster hero, Celtchar mac Uitechair. Sanctuaries and necropoli (cemetaries), including the seven monuments in Moanmore East and Doon Hill, were theirs. A standing stone in Masonbrook overlooking Rathsonny fort and another on the heights of Earlspark overlooking Brickloch, are but two of the many monuments here. After Maine Mor's 5th century invasion this archaic landscape around Loughrea became part of Hy Many. The Dal ngChruithne revived their fortunes by defeating the Ui Maine in this area E of

Loughrea in 802 A.D. with the aid of their kinsmen, the Sogan of Tiaquin.

The set of ramparts then ran W across Earlsparke along the steep S slopes of Mount Pleasant and Knockanima to the SE side of Loughrea Lake. The Sli Dala highway (an ancient road) formed the Mount Pleasant/Knockanima boundary, climbing up from the E side of Loughrea town to the heights above the lake. This road swung ESE forming the Earlspark boundary with Mountpleasant and Moanmore West, heading for Portumna and on to Roscrea. A branch road divided from the Sli Dala here, heading S over Aille along the E slopes of the Aughta Mountain. At the NE end of the lake another branch headed S along the W side of Aughta Mt. Joe Dunne pointed out the different routes of these ancient roads.

Loughrea lake was a sacred lake in Queen Medb's day. Aillil, keeper of the lake, became Medb's consort and King of Connaught. Medb had a Royal Crannog on the lake. While bathing there, she was finally assassinated at the instigation of Connor Mac Nessa.

The set of ramparts skirted Loughrea's S lakeshore as those of the previous set did the N shoreline. Transverse ramparts still run radial-like on the slopes above the lake. A rare example of defensive embankment with outer fosse facing uphill forms the field boundary just below the front of Shield's house in Grange on the slope down to the SE side of the lake. It ran W along the heights and slope overlooking the lake through Grange and Grange Park. Numerous forts studded the rampart route along this belt above the lake. Twelve are shown on the 1840 O.S. map but only eight on the 1933 map. Most of these have now disappeared. Inexplicably, the surviving vestiges of ramparts overlooking the lake in Grange and Curheen are not marked on the O.S. map. The set of ramparts bent rimlike, accommodating itself to the curve of the lake's S shoreline continuing W through Curheen and Glenaclara East and West. Vestiges survive in Curheen and Cuscarrick. Sheamus Callanan of Knockauncaura guided the reconstruction from

Figure 3.10. Ramparts such as this run for long distances across the countryside.

Loughrea W to Kilchreest and beyond, through Glennaclara, Aille, Cahercrea East/West, the 2 Ballingarrys, Gortnamacan, and Killaspugmoylan, along the marshlands of the W-running stretch of Kilchreest river to its W zig-zag curve where it swung away SW. He pointed out hard to find traces of ramparts in Kilchreest and Gortnamacan along the river whose floodlands were utilized to a supreme degree. Sections of the river were banked up and floodgates fitted all the way to the S end of Ballygarraun to induce wide- spread flooding for defensive purposes. The 1840 O.S. map shows dyke vestiges on both sides of the river in Kilchreest, Gortnamacan, Ballingarry and Ballygarraun. Some became rights-of-way and later formed local roads. Some were levelled before the revised 1933 O.S. map was published and were not recorded. The 1840 O.S. map shows the complicated web-like nature and interrelation of onward and transverse ramparts giving a fine example of the confusing and classic nature of the Belgic defensive system.

The rampart set swung NW from the N tip of Roxborough through Pollnashinnagh (*Poll na sonnagh* = rampart bog), Ballyshea, Caherskeehaun, Creggmulgreny and Lakyle. They skirted Lough Kinlea (Cinn Chlaidh = Dyke head) and Lough Burke at the N tip of Ballyshea and Knocknamanagh. They traversed Carheen and Emlagh, skirting Killogilleen and Caher Gal. They ran through Ballynascragh, Ballylin West and Bauttagh, skirting Ballymanagh and Kilquain. They sped through Ballynamannin and Mannin. Ballynamannin, Ballymanagh and Knocknamanagh proclaim the presence of the Manapii Celtic tribe (*Manach*). The ramparts raced through the W side of Roo (rampart), in which there is a Lisdoo (fort of the rampart), Aggard Beg with its Lisroo (rampart fort) and Lissindragan (*dragan* refers to the dyke) right on the lines of fortification. Skirting Aggard More, they cut their way through to the marshlands to the edge of the original Rahasane Lake and the Dooeyertha (Dughiortach/*Dur* or Dunkellin) River in Carrigeen East and West. Across the lake and river stood the ramparts of the outer ward of the Athenry oppidum in line with the latter as they ran NNW to Carnmore and Lisheenavalla, thus welding the two sister oppida of Turoe Knocknadala and Athenry into one joint oppidum.

We will now go back and examine the beginning of the Athenry oppidum.

Location map for the area around Athenry

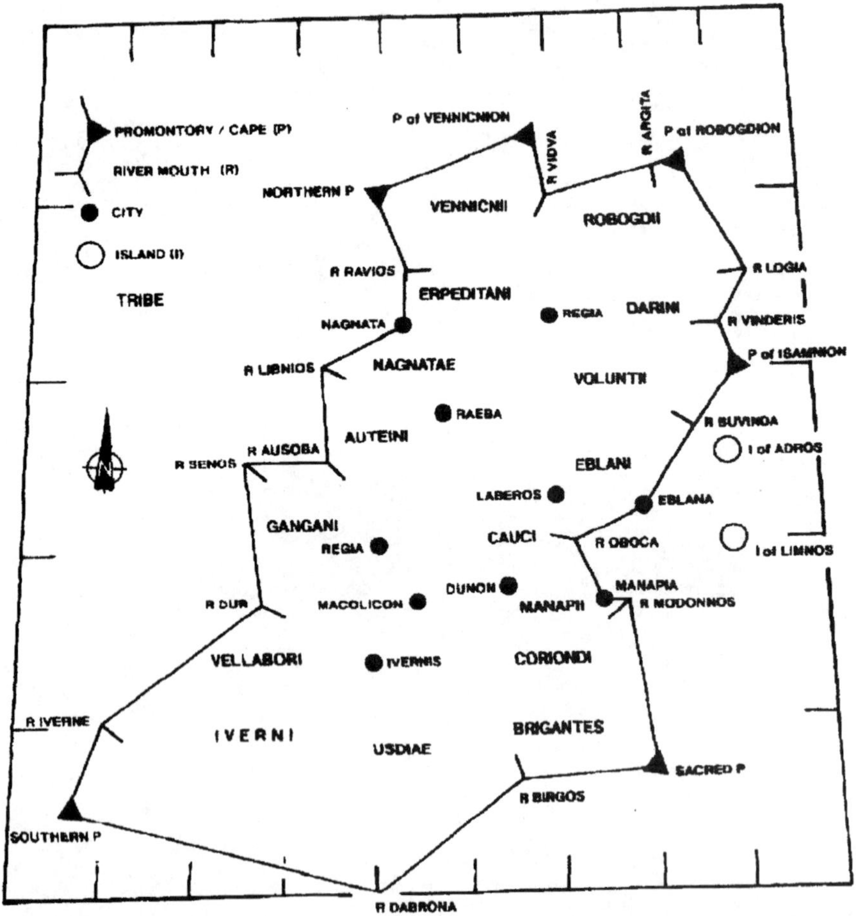

Figure 4.1. The map of Ireland in the 1st/2nd century as generally drawn from the co-ordinates left by the geographer Ptolemy of Alexandria. On the west coast this has been misinterpreted, as described in the text.

Chapter 4.

Linear embankments around Athenry in County Galway

Athenry is well recognised as a Norman town and the evidence is there to be seen. What few people realise is that Athenry has a history dating much further back than the Normans, in fact back to the 1st century A.D. In Ptolemy's 1st/2nd century record of Ireland the name 'Auteinri' appears in the place where Athenry is today (Figure 4.1). Ptolemy's 'Auteinri' is a perfect Greek rendition of the ancient name of Athenry, Ait an Ri, as it was pronounced. It strongly corroborates what archaic legend tells about Athenry and its surrounding area. In ancient Dindsenchas documents the plain around Athenry is called 'Magh Muc Dhruime', although deliberately mis-spelt as Magh Mucrimme by pseudo-historians. This mythological name, *Magh Muc Dhruime*' (Plain of the Magic Boar's Dyke) corresponds to the mythological name of the Great Wall of Ulster erected by the men of Ulster against the Fir Belg of Connaught, namely, 'Cladh na Muice Duibhe' (The Magic Black Boar's Dyke or The Black Pig's Dyke). Magh Muc Dhruime got its name from the mythological description of the erection of the linear embankments, of the inner and outer wards surrounding Athenry. These embankments are similar to those around Turoe/Knocknadala, and were allegedly thrown up by the tusks of the mythological wild boar chased by Queen Medb around the plain of Athenry. This tale is the mythological description of the erection of these earthen embankments around her kingdom and it is clear from the tale that Queen Medb had her palace at the core of these linear embankments, namely, at Athenry. This was subtly suppressed by the pseudo-historians for their own ulterior motives as told later. Elderly countryfolk refer to these earthen ramparts around the town walls, as the 'Inner Dun' of Athenry, and their corresponding sets of outer ward ramparts as the Great

Duns, Outer Duns, Muc Dhruim, Muc Cladh or Dun Cladh (= encircling defensive dyke).

Sixty years ago, long before any author had written about the unique nature of the Belgic defensive system in SE England or on the Continent, one of these local elders, Martin Finnerty, wrote a short book entitled 'Punan Arsa' ('Archaic History Notes'). This is a series of notes on aspects of the early history of the area ages before the coming of the Normans. He described an extensive stretch of the 'Great Dun' which ran all around Athenry, four miles out from the town. Although not using the archaeological terms of 'Inner Ward' and 'Outer Ward' he showed that old people of his day were aware that this was an outer expansion of the ramparts around the medieval Norman walls of the town. He recorded the seven mile stretch of rampart from Greyford in Kiltulla to Roundfield in Monivea. He described a massive section in Roundfield on which he played as a boy over one hundred years ago. Only surface traces of this rampart survive today. His ground-breaking document lies unknown and unrecognised in the National Library, Dublin.

This present reconstruction, then, cannot claim to be original. It owes much to oral history handed down by the countryfolk, such as Mr. Finnerty, covering a wide area around Athenry and, indeed, the whole of Galway and beyond. It is not the intention here to present a definitive reconstruction of the Athenry oppidum, but rather to show how it stands as a Belgic oppidum in its own right in relation to the Turoe/Knocknadala oppidum, how it merges with, and is eventually overtaken by the it. Full credit is due to Martin Finnerty for putting the Athenry Belgic oppidum on record long before similar Belgic Capitals with their unique defensive systems in SE Britain and in Belgic areas of the Continent were placed on record. His documentation of the entire Athenry oppidum is incomplete and rudimentary. He was actively gathering more information and carrying out research to produce a fuller account, not only of the Athenry oppidum, but of that of Turoe/ Knocknadala also, until he passed away. He was ahead of his time in recording the existence of the unique Fir Belg defensive system, however simply expressed.

The Athenry/Rath Cruachu/Magh Muc Dhruime oppidum

Today only a few are aware of Athenry's ancient names, Rath Cruacha/Cruachan and Magh Muc Dhruime (Plain of the Magic Pig's Dyke). Magh Muc Dhruime referred to the whole area within the linear embankment outer ward surrounding Athenry. Although Rath Cruacha/Cruachan is still known by elderly people as the name of several forts surrounding the town of Athenry and as the ancient name of the area in general, the significance of these names and their location are no longer appreciated nor understood. They have been hijacked and transported to an entirely different scenario, i.e. Rath Croghan of Roscommon. When the Ui Briuin Kings of Connaught transferred Queen Medb from Rath Cruachu of Athenry to Rath Croghan of Roscommon, they swapped the name of Rath Cruacha/Cruachan of Athenry for Rath Croghan of Roscommon. Rath Croghan was never known as Rath Cruacha/Cruachan until then. Even today, elderly residents within a fifteen mile radius of Rath Croghan of Roscommon fervently insist that Rath Croghan is not, and never was, known as Rath Cruacha/Cruachan. The name Rath Croghan as pronounced by the inhabitants of that vicinity of Roscommon sounds exactly like Rath Cruthin. This makes perfect sense as it was the stronghold of the aboriginal Cruthin in the West of Ireland and, as such, was known as Rath Cruthin.

Rath Cruacha of Athenry was a Fir Belg stronghold in Queen Medb's day. Though smaller than the Turoe/Knocknadala oppidum, the names of Rath Cruacha and Magh Muc Dhruime are better known in legend and history than those of Turoe or Knocknadala, which were suppressed. Earthen ramparts of Athenry's inner ward are now wrongly ascribed to the Normans. They were erected 1200 years before the coming of the Normans. It is widely thought that Athenry had no history before the Normans came and erected town walls and ramparts. Elder countryfolk

Figure 4.2. Earthen ramparts surround the walled town of Athenry. This section is on the east of the town.

Figure 4.3. A section of ramparts on the northwest , just outside the town walls. As can be seen, part of the wall was built on the rampart.

confound this 'official doctrine' by claiming that the earthen ramparts of both the inner and outer wards surrounding the town were erected by the ancient Fir Belg.

Inner ward of Athenry/Rath Cruachu/Magh Muc Dhruim

Athenry's inner-ward ramparts lie along the medieval walls, which form Athenry's boundary with Gorteenacra and Knockanglass on the E side. Baunmore (cited as the place where Medb held war-games in a pre-battle Assembly at Rath Cruacha) and Prospect are on the S. Raheen and Ballygarraun South on the W, and Culairbaun and Caheroyan on the N (see Location map, Figures 4.2 and 4.3). Caheroyan (*Caher Roighan*, Old Irish form of *Righain*, Queen's Palace), makes its own profound royal statement which Athenry has not taken seriously to its utter detriment. Note, Athenry is not Ahenry (Atha na Ri = Ford of the Kings, a false translation of late medieval origin). The 'At' of Athenry, as it has always been pronounced, is the Irish word 'ait' meaning place of residence or home. Athenry is the 'AUTEINRI' of Ptolemy's Irish record. The Greek 'Auteinri' like 'Athenry' when broken into its separate components spells Ait en ri = 'Ait an Ri', Residence, Seat or **Home of the Kings**. As it was in Ptolemy's 2nd century, so it is today.

The name of the original site of Caheroyn, site of the Norman Castle, speaks of a Queen (Old Irish *Roighan*), not a King. Who was this Queen? As this book grows on the reader, so will this Iron Age Queen of Rath Cruacha of Athenry. Athenry, not the Norman town, but the Belgic oppidum, was her father's creation, not that of the medieval Normans. When her father, Fir Belg Overking (Ri Temhro) Eochu Ferach Mhor (also called Eochu Felech), who lived at Turoe/Knocknadala, commissioned Tinne mac Conrach to erect ramparts around Rath Cruacha it was not at Rath Croghan of Roscommon that Tinne and his men erected the ramparts but at Rath Cruacha of Athenry, Co Galway. Tinne was King of the Belgic Domnann sept who were positioned in small buffer states on the W and N flanks of the outer ward defence line of the expanding

Belgic Oppidum. In repayment for so swiftly erecting the earthen fortifications around Rath Cruacha, the Fir Belg Overking made Tinne first King of Rath Cruacha. *'Cath Boinne'* relates that Tinne became embroiled in conflicts and so was demoted in favour of Medb, a daughter of Eochu Felech (Ri Temhro). From Rath Cruacha she ruled not only over the Fir Belg of Connaught (called Ol nEgmacht at that time) but over the combined Fir Belg forces of Ireland and led them against the Cruthin of the North. Tinne remained at his stronghold, Caher Tinne (now Cahertinny townland), some 2 miles NW of Loughrea at the N side of the Ardnadooman, (corrupt form of Ard na Domnann = height of the Domnann), but continued to be a frequent guest and paramour of Medb at Rath Cruacha.

Athenry/Rath Cruachu/Magh Muc Dhruim oppidum outer ward

From Mount Shaw fort in Ballygarraun, the stone-core ramparts of the outer ward reverted to earth as they ran through Raheen, Culairbaun and Caheroyan. They swung SSE straddling the Gorteenacra/Kingsland North boundary into Kingsland South. They swung SSW straddling the latter's boundary with Rahard through Baunmore to the N tip of Boyhill. They swung NW through Gortnahown and NE Newford back to Mount Shaw. Several forts hugging this line of ramparts, like that in Athenry cemetery, and another adjoining it (now demolished), were known to elders as Rath Cruacha.

First extension of the outer ward of Athenry/Rath Cruachu/ Magh Muc Dhruim oppidum.

An extension of this outer ward ran ENE from Moanbaun, a mile NW of Athenry, through Park and ran astride Ballydavid South from its N boundary to its S road. From Carrowntobber West it swung (SSE) through Kilcornan, Pollacappul, and the E of Kingsland (*'Ait na Ri'*) to Blean. In the S of Blean it swung SW (WSW) straddling both the road and boundary with Backpark.

Rampart traces survive along this roadside and in Kilcornan and Pollacappul. The set of ramparts swept through Rahard. It embraced Loobroo (Lugh Bru = the god Lugh's Otherworld Abode) and Boyhill, 'Cnoc Boidbh = the god Bodbh's Otherworld Abode, also known as Sidh Cruacha, Sidh Connaught or Relig na Ri (Figure 4.4). Surface traces of several tumuli (burial mounds) survive there, as shown in Figure 4.5. A vestige (part of) of the main rampart and a transverse rampart, both of stone, also survive here. The Bottom (*an Tonnagh* = rampart) road was partly built on a levelled rampart. The ramparts skirted Tober Conaille (Conaill, whose palace stood in the next townland, was King of Connaught), and the N of Mountain South (*Cruachan Theas*). This is one of the 4 Cruacha around Rath Cruacha of Athenry which are designated after the cardinal points of the compass and have been wrongly anglicised as 'Mountain'. Across the Craughwell road in the NE tip of Clamper (Cladh an Farra = Sentry dyke) are the ruins of a double-ringed stone caher known as Clochar Chonaille, the palace of Conaill, King of Connaught with whom Conn of the Hundred Battles, a later King of Connaught, spent his fosterage. This Royal Palace was one of the finest stone forts in Ireland before it was badly defaced in recent times by a bulldozer. Following the banks of the Escir river, the rampart set straddled the Clamperpark/ Turloughalanger boundary to that of Cahereenascovoge. There it swung NNW through Castleturvin, Millpark and Mulpit. It traversed the NE half of Cloran and its border with Newford. It crossed Ballygarraun West and North, and the E protrusion of Rathmorrissy. Several forts flanked this line of fortification. It straddled Pollnagroagh/Ballygaraun North boundary back to its Moanbaun starting point.

Second extension of Athenry/Rath Cruachu/Magh Muc Dhruim oppidum outer ward

Reconstruction of a 2nd extension of the outer ward starts almost 3 miles NW of Athenry. It ran from the Moor/Deerpark boundary through Tobernavean (Well of the Fianna), Carnaun, Pollagh and Fahysvillage, astride the latter's boundary with Mountain North

Figure 4.4. Boyhill, just south east of Athenry, showing one of several denuded stone cairns in the grove of trees on the left.

Figure 4.5. Surface traces of stone cairns, such as this, are found in Boyhill.

and Castle Ellen. It sped ENE through Loughaunenaghan, Skeaghadcrcen and Knockbaun into Montpelier. In Carrowantobber East it swung SSE through Gregabbey and Kilcornan, and ran astride the Gloves West/Gloves Middle boundary. This Irish name of the several Gloves townlands is 'Na gChlaidhbhe' = the Dykes ('double-ditches' locally), a name which emphatically spells out the ancient character of this area of strategic defence. The prevalence of so many Gloves townlands is because the Athenry and Turoe/Knocknadala oppida outer wards left numerous vestiges behind here.

The ramparts ran through Killascaul along its boundary with Backpark and Deerpark. The forest of Coill A Scail was extensive in ancient times. It bears the name of the Celtic god *'A Scail'*, also known as Cian. Cruchain Cian, a gravel hill recently demolished, is closeby. The drained Lochaun Escir and Escir river forming the N boundary of Killascaul were known as Loch A Scail and Abha A Scail, respectively. The upper River Scheld in Belgium in the lands of the Belgic Atrebates and Morini was known as A Scail (*A Scaul*) as shown on Roman maps of the period. From there the Atrebates (Aitreabh) and Morini (Mac Morna) came to the W of Ireland via the SE of England and probably brought these names with them.

The ramparts skirted Esker and Rathgorgin along the slopes of Cruacaunanaine (*Cruachan Aine* = Otherworld Abode of the Celtic goddess Aine). The presence of the sacred shrines of Cian (*A Scail*) and Aine right on the boundary between the two oppida is significant. In Continental Europe sacred shrines of Celtic gods and godesses often stood on tribal boundaries which were also places of fairs and markets. Right on the boundary between the two oppida at the S foot of Dominic's Hill there is a raised rectangular earthen embankment with outer fosse. Its significance is unknown. The possibility of a Celtic Temple on the boundary where the two oppida marched together cannot be ruled out.

The ramparts ran through the W of Rathgorgin (Ailech Gorgin), Fort of Gorgin, also known as Corc. When Corc came to Ireland he was made King of the tribes left behind by Lughaid Mac Con. When Cormac Mac Art reclaimed the throne from Lughaid Mac Con, Corc was forced to flee with his rebellious followers. He went forth to found Cashel, the later royal Capital of Munster.

The set of ramparts straddled Tallyhoe Cross Roads skirting the E side of Moyode (Magh Fot or Fotad) of ancient fame, plain of the Fotadini. The Fotadini warriors, known in Ireland as the Fotads, also came in the war band of Lughaid Mac Con from the Votadini tribe in the SE of Scotland. The Fotad palace, Clochar na ngFhotad ('Clochar na ngFhot'), which was named after them stood in Moyode (Magh Fhot). Two sites of this palace remain, one destroyed, one remaining in the form of a stone caher, but neither of which are shown on the O.S. map, are still pointed out by elders from either side of this vast townland.

From Moyode the ramparts, with stone cores (see Figure 4.6), ran through Slieveroe (Sli Bhrugh = road of the Otherworld Abode of the gods) and Rockfield, cutting through the N of Knockatoor and Cahercrin into Moneyteig. They swung NNW parallel to the NS stretch of the Escir/Lavally river which they crossed. They commandeered Caherfin Esker. From Greethill they ran astride the

Figure 4.6. In Myode, a townland with abundant stone, earthen ramparts are replaced with stone ramparts.

Gortroe (Rampart park)/Derrydonnell More boundary and the Rathmorrissy/Tobberoe/Caraunduff boundary. From the Caherbriskaun/Knocknacreeva boundary the set of ramparts swung E through Pollagooil and Deerpark astride their boundaries with Castleambert on the E side and Moor and Ballinloughan on the W. They thus returned back to the starting point at Moor/Deerpark/Tobernavean.

At the centre of Carnaun (cairn tumulus with traces of satelite cairns) are the ruins of what was called King John's Castle by some, and by others Clogher Goill, residence of Goll Mac Morna, leader of Connaught's Fir Belg Fianna fighting men who manned numerous sections of the ramparts. They were given lands for their services. 'Clogher' is anglicised on the O.S. map as 'Convent', confused with modern Irish 'clocher'. Old Irish 'Cloghar' meant palace or princely residence. The Diocese of Cloghar was named after Clochar Damhan = palace of Damhan, an Airghiallian 5/6th century king, although in later times a convent was built on the site. In the Carnaun surroundings there are a number of local placenames referring to the Fianna, such as Tobernavean (Tober na bhFhianna).

Third extension of the Athenry/Rath Cruachu /Magh Muc Dhruim oppidum outer ward

Working from the point in Rathgorgin where the Athenry outer ward extended out to meet the Turoe/Knocknadala outer ward, the linear embankments of a third and final outer ward extension of the Athenry oppidum ran WSW through Moyode Demesne. They straddled the area between the Moyode road and the Doreythra (*Dughiortach/Dur/Dunkellin*) River to both of which they ran parallel. In Moyode Demesne stone ramparts were in a tumbled down state until the 19th century, when the Persses of Moyode paid men one penny per day to rebuild them as wide straight up walls and fill in the fosse. The set continued through Ballywinna and Temple Martin skirting their borders with Pollnabanny and Ballymore/Craughwell. They ran astride the

Parkroe (rampart field)/Shanbally/Killeely and Ballywullagh/ Crinnage boundaries to the edge of the former extensive Rahashan lake. This was remembered until the 1950's as 'Dur Loch'. This ancient name is corroborated by the name of an ancient road, Boher Doorabhaigh (Boher Diur Abha = Diur River road) running NW from the W edge of this lake to an inner harbour in Oran Beg. Irish speakers 50 years ago insisted Boher Diur Abhaigh was the correct name of the road, not Boherdooroghy as it is on O.S. maps. There is abundant evidence that the sea was much higher in the early centuries than it is at present. At high tide it flowed right into Dur Loch which served as an inner harbour. Traces of landing places remain. Sea shells on the undersides of large rocks in the area can only be explained by the presence of the sea in former times. British experts have shown that the sea was 7 to 10 feet higher in the Roman period (2nd to 5th century A.D.) than it is today.

Above the N shores of the former lake in Ballynagran, Caherfurvaus, Carheenadivane, Caheradine and Kileeneen Beg and Mor stand a line of strong forts, their main concentration being close to a natural landing place for boats. A nearby fort bears the significant name of Caherfurvaus, Caher Foir Mhas = broad hill ('Mas') fort of the sailor's shelter ('Foir' = sailors haven whilst waiting for high tide to set sail). The same Old Irish word 'Foir' is found again in nearby Foorkill (Foir Choill = wood of the sailor's haven), wrongly anglicised as Coldwood, the Old Irish *'foir/faire'* being confounded with modern Irish word *'fuar'*. The 2nd-5th century sea levels also reached Foir Choill along the Clarenbridge river channel. Seashells are found on the undersides of rocks there as at Rahashan. Besides the Coill (wood) of Foorkill, there was a Fochoill. The entire forested area along the lower Dunkellin (Dughiortach/Dur) River is referred to in ancient texts as Fochoill, Ochaill or Achaill Oichne. St. Patrick, fleeing from his slave master on Slieve Mish in Ulster guided by an angel, took ship for home from here. In his 'Confessions' he named the place where he waited for his ship as 'Foculut'. He stated that it was not on, but close to, the western sea shore and that it was 200 Roman miles from his place of slavery. This inner harbour of Ath Cliath Magh Ri beside Fochoill (Old Irish *Fochoillut*) is 200 Roman miles from Slieve Mis, unlike the Kilalla harbour site of the N Mayo coast, the so-called 'traditional site of Foclut' according to the pseudo-historic record of Tirechain. Ath Cliath Magh Ri, 2 miles W of Rahashan and Coldwood, was the illustrious "sea-port of ancient Ireland where all ships sailing the high seas called in." Coldwood Turloch, like Rahashan, was a large lake. All that remains after the new deeper channel was cut is Pollnacirca birdhaven. Traces of landing sites survive at the original inner E end of the Foorkill/ Coldwood lake.

From the NW end of Rahashan the rampart set ran NNW from Kileeneen through Caherbulligin, Creganantaa and Brochagh to Coldwood. Vestiges of watered-down ramparts survive on the N edge of Coldwood lake (Figure 4.7). They crossed the sea-washed plain of Moyveela (sea plain), Shantallow, Derrydonnell Beg, Mountain West (Cruachan Thiar), and Derrydonnell North. The Dublin/Galway railway line crosses a long 25 ft. wide stretch of rampart near Derrydonnell Crossroads. Transverse ramparts survive in the Mountain West Forestry plantation. The set ran

Figure 4.7. Vestiges of parallel ramparts (one in the foreground and one in the centre of the picture) at the edge of Cooldwood's drained Pollnacirca lake, into which the sea formerly flowed at high tide.

NNW astride the Mountain West/Derrydonnell North boundary. Lisroughan (rampart fort) mans this line S of the railway line in Mountain West. It straddled the Carnmore E/W boundary to Lisheenavalla and Island, then swung ENE.

The set of ramparts skirted the extensive ENE lower Cashla protrusion passing through Ballymoneen and Grange East. It crossed Peakroe, Caherteemore South, Sheeaun Park, Cossaun, Saintellen, Castle Ellen and Caraun. Ned Ruane of Monivea Road, with a store of oral history, claimed that Saintellen/Castle Ellen are derived from the same ancient name, a corruption of Sidh nEthelann (Other-world Abode of the Celtic goddess Ethlenn, mother of Lugh, most famous of all the Celtic gods). The rampart set pressed on through Laughaunenaghan, Hundred Acres, Knockbrack and Costello Park to the Pollavullaun/Corrabaun boundary with Roundfield. It ran from Roundfield in Monivea SSE astride the Corrabaun/Pollavullaun, Knocbrack/Mount Garret, Tisaxon/Templemoyle, Bingarra/Cappaghmoyle, and the Lisdoran/Toorkeel/Glenagloghaun boundarries on to Ballyboggan, Cloonbrusk, Gloves East, Middle and South and their boundary with Carnakelly North/South to the Curraghmore/Greyford boundary. This is the rampart noted by Martin Finnerty in his "Punan Arsa" 60 years ago where he described this outer ward "Great Dun" surrounding the Athenry oppidum. This ran from Greyford in Kiltulla through 7 miles of countryside to Roundfield in Monivea and thence W. It swung round the W side of Athenry and thence back again to the starting point in Greyford (Rath Chlaidh). This rampart is clearly shown on the 1933 O.S. maps, as shown in Figure 4.8. Sadly, very little remains of this rampart. In the 1930's Martin Finnerty told of "the ancient dun that is known to exist to the present day and runs alongside of the town" of which the Greyford/Roundfield section 5 miles out from the town is part of an outer expansion. More details were supplied by Ned Ruane of Monivea on the ramparts running W through Mountbrowne and Carheenlea to the Clare river before these were levelled out. He referred to this as "*Dun Cladh*" of Athenry.

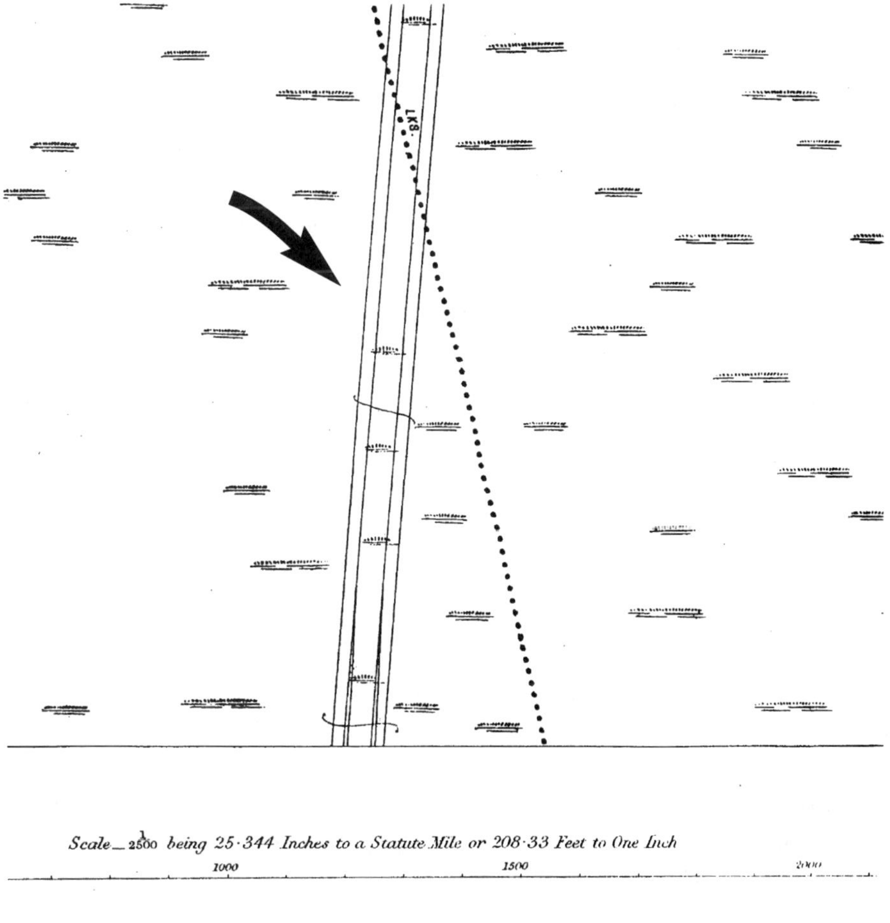

Figure 4.8. The ramparts so clearly and vividly described by Martin Finnerty in the 1930's in his book 'Punan Arsa', as shown on the 1894 O.S. map. Using the scale on the map, an accurate width can be obtained for the rampart (20 metres). Unfortunately, this rampart, like so many others, has been totally destroyed. Map courtesy of Trinity College Dublin.

Mr. Finnerty claimed these outer ward ramparts like those immediately outside Athenry's Norman walls were attributed by the old people 100 years ago, not to the Normans, but to the Fir Belg. He wrote: "authorities have written that this town had no civil history before the arrival of the Normans". Against this he put forward the tradition of the people 100 years ago: "Let it well be borne in mind that Athenry had a name, an ancient name with a storied past, and a background of antiquity when Ptolemy wrote about it" in the 2nd century A.D. Had it not, Athenry would not have figured on his geographical record as Αυτεινρι (Ait an Ri = Home of the Kings). Finnerty's *'Punan Arsa'* preserves a valuable description of a rampart in Roundfield as it was over 100 years ago. His record is the more precious since it has since been obliterated, and there is no other rampart in such excellent condition with fosse intact. He states that as he stood on top of a section of rampart in Roundfield as a boy "a feeling of awe came over me on looking down" He describes its construction as it appeared to him: 10 feet wide on top and more than 10 feet high above ground level with a sloping stone face on either side. The deep fosse added another 10 feet to its outer face presenting a formidable high obstacle to an invader. It was erected by cutting "a canal at least 10 feet deep" on the outer side through "a carefully selected ridge of country". Many old people who knew of similar sections describe the linear embankments in similar fashion.

Mr. Finnerty noted that the "Great Dun", as locals called the outer ward of the Athenry Belgic defensive system, corroborated Ptolemy's record. He declared that the "Dun of Athenry is evidence of the fact that it was the work of the Fir Belg (the Belgae); there was no other tribe...to do such marvellous work", as has since been corroborated by British and Continental archaeologists who have elaborated on this unique type of defensive system developed by the Belgae alone. The Ulster Cruthin built the 'Great Wall of Ulster' known as the 'Black Pig's Dyke'. This, however, was in answer to the advance of the Fir Belg using an idea learnt from them for their defence, but of a different style. The Cruthin relied on a single massive rampart unlike the Fir Belg. It ran from one coastline to another cutting off the area occupied by the Cruthin rather than surrounding their Capital of Emain Macha. The parallel Muclahs (*Muc Cladh*) at Rath Croghan of Roscommon are more like the Fir Belg ramparts. Mr. Finnerty was way ahead of his time in recognising that the Belgae alone, of all Celtic tribes, developed this unique type of elaborate defensive system encircling their capitals, a fact that has only been discovered and recorded in Britain and the Continent in the last 30 years. This rampart set ran through Gloves East, where a short almost intact section of original rampart still survives.

The embankments surrounding Athenry and forming the outer extensions are shown in Figure 4.9.

The relationship between the Athenry Oppidum and the Turoe/Knocknadala oppidum is shown in Figure 4.10.

This ends the expansion of the Athenry/Magh Muc Dhruim oppidum in its own right as it will now fully merge into one joint oppidum with the Turoe/Knocknadala oppidum. From this point on the expansion of the two oppida will be reconstructed as one joint oppidum.

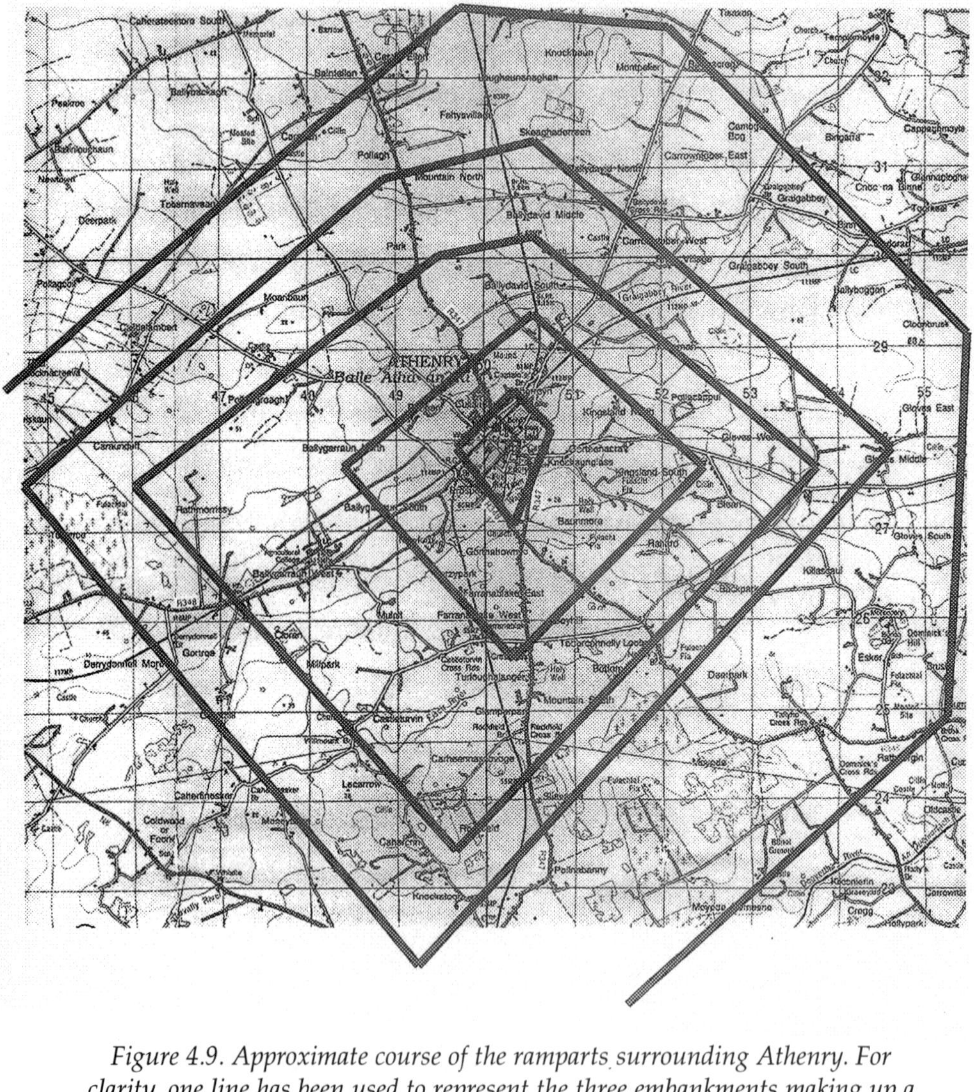

Figure 4.9. Approximate course of the ramparts surrounding Athenry. For clarity, one line has been used to represent the three embankments making up a set. (Map courtesy of Ordnance Survey Ireland, Permit number MP 0001903).

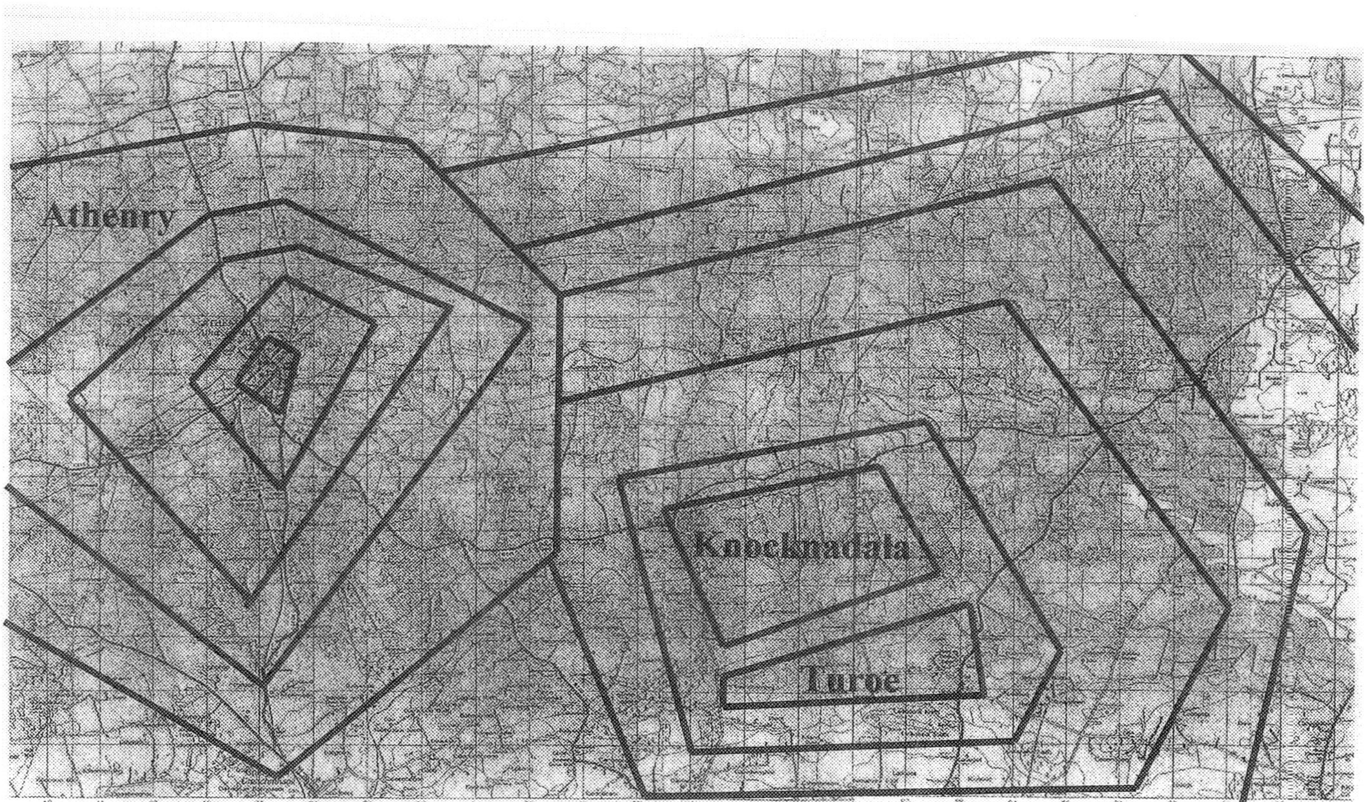

Figure 4.10. The relationship between the Athenry Oppidum and the Turoe/Knocknadala oppidum.

Figure 5.1. Vestige of a linear embankment in the Roundfield area.

Figure 5.2. The burial (cairn) mound of Diarmait Mac Morna, brother of Goll Mac Morna, the leader of the Fianna in this area. Cairn Dhearmait has given its name to the townland, Corrantarramud.

Chapter 5

Extensions of the linear embankments around Athenry/Turoe

As we have seen there was one oppidum at Turoe/Knocknadala and one at Athenry (Rath Cruacha) that were joining along the Greyford/Gloves/Clonbrusk/Moinviea area. Rather than compete with each other for territory, the occupiers of these oppida decided to join forces and therefore, have a greater impact in gaining territory from the Cruthin. And so various extensions of this joint oppidum can be reconstructed.

First extension of the joint Turoe/Knocknadala/Athenry/Rath Cruacha oppidum outer ward

Reconstruction of the first extension of the outer ward of the joint Turoe/Knocknadala/Athenry/Rath Cruacha oppidum commences from its NW corner on the N banks of the Clare river in Rooaunmore (Rampart set) and Kiniska (Ceann Isca = head of Turloughmore lake) to the W of Claregalway. The set of linear embankments ran E across the Clare river, straddling the Lacaghbeg/Cahershilleeny, Coolaran/Grange East, Carheenlea/Caherateemore and Rathfee/Glenmore boundaries on to Killsceagh. Cahermore fort stood on a commanding height with the slopes providing a strong position. The set ran along the Gortavaura/Mountbrown boundary to Farravàun, skirting Bellville Demesne, straddling the Corralea/Roundfield (Figure 5.1) and the Pollavullaun/Corrabaun boundaries with Pollboy. It crossed Cartron ('Cnoc a Farra' = sentry hill) and the tops of Glennagloghan South, Glennaslat and Corrantarramud (corruption of Carn Dhearmait = tumulus of Diarmait Mac Morna, brother of Goll Mac Morna, the leader of the Mac Morna Fianna defending this front-line defence). His burial mound sits on the hill beside the Newcastle road near its W end (Figure 5.2). The line of

Figure 5.3. Vestige of a linear embankment in the Kilconnell area.

ramparts ran straight on along Corrantarramud (Carn Dheamuid)/Glennaslat boundary, Lenamore, Tiaquin and the Cormacuagh/Derroogh boundary. It crossed Shanballymore and Ballyglass where the dykes were known as *'na curraigh'* heading towards Gurteen church. It straddled the Cloonbornia/Gurteen, Gortnahultra/Creeraun, and Corskeagh/Kinreask borders to Cave, the Ballinlough/Carrownakelly boundary, Carrownea, Tullawicky, Liscune and Gortyroyan townlands to Callow Lough (*Loch Acalla*). There the Curha Dyke swung SSE through Cartrondoogan (rampart quarter) beside the railway line. It ran on to Kilconnell through Ballyglass, straddling the boundaries of Abbeyfield and Corraneena, into Gortadeegan (Dyke field; *'dig'*, *'deegan'*, and *'dickeen'* are corrupt Gaelicized forms of 'dyke' common in Connaught and Ulster). From Kilconnell (Figure 5.3) this set of ramparts ran parallel to the W side of the first half of the Kilconnell/Aughrim road and thence on for 12 miles in a more or less straight line to Killimor. Rampart vestiges remain along that section of the Kilconnell/Aughrim road, which was partly laid out along the fosse through Ballinderry, Killareeny and Ardross. The set skirted Garrymore as it ran through Lavagh, Newcastle, Moat, Cloghagalla, Oatfield and Ballynamuddagh. *Muddagh* is a gaelicized rendering of mootes, the word commonly used in SE Galway for the ramparts (double-ditches elsewhere). From Lurgan Great it ran on for 4 miles to Lurgan More. From Derrew and Lisdelingy it cut through Derrysiskal, Corry, Rusheeny and the Ahanduff townlands to Moneenaveena (bog of the Fir Belg Fianna). Numerous vestiges of linear and transverse embankments survive in these townlands. Local tradition claimed that the old name of the Kilcrow river was Abha an Doo (Dyke River, corrupt Ahanduff). Some stretches of the river were banked up into dyke ramparts in the classic Belgic defensive style. Legendary history links Killimor with the Fianna and Fir Belg. It is still known as Killimor Bolg (Killimor of the Fir Belg), to distinguish it from Killimordaly. This was a well-known stronghold of the Fir Belg fighting men who held the Cruthin of Ulster and of E and NE Connaught at bay along this defence line for a long period before they were eventually pushed back behind the Suck river. The Fianna left their name on Moneenaveena. Ardanena at Killimor was a Fir Belg power centre and gathering place.

On the S side of Killimor the rampart set swung W along the Garryduff (*-doo* = dykeland)/Treananearla boundary through Killeen, Cloonacusha, Tynagh, Derryfrench, Bracklagh, Grange, Ranamackan, Limehill, Leitrim Beg/More, Dalystown, Ballyknock, Knockash, Knockroe, Bookeen, Curhoor and on the heights above the Killeenadeema/Aille road. It ran on through Burroge, Kylegarriff, Ballynacurragh (Dyke townland), Gortnamannagh East/West (townland of the Manapi), Gortnagleav (Dyke townland), Cloghaun, Cregganore and Castleboy. The defence of the forested S flank of this outer ward was enhanced by extensive boglands on the heights of the Aughta mountains. Belgic engineers directing the erection and route of these outer ward fortifications utilized all natural defences along the route, such as forest, bogs, marsh-lands, hillslopes, mountains, lakes and streams, incorporating them into the defensive system.

The rampart set swung NNW through Pollacurra, Ballinrooaun (rampart townland), Isertkelly, Moneen East, Furzeypark/ Fiddaun, Shantallow/Ballyglass East, Cloghroak and the NE tips of Ballybaun and Toberacreggaun. It ran astride the boundaries of Gortroe and Liskevin, Parkalaughan and Fawnarevagh, Lavallyconnor/Pollnagarragh West and Carrowmaneen (of the Manapi), Caheraloggy West and Parkroe, and on through Garraun and Rinn. It crossed the Dunkellin river, straddling the narrow neck of land between the former Rahasane and Dunkellin Turlough lakes linked by a narrow channel. It sped through Caherapheepa manned by four strong stone forts, along the Fahymactibbot/Roevehagh (rampart set)/Killeeneen More/ Ballygarriff road. This ancient road known as *Boher Dur Abhaidh* was built on the earlier rampart. This archaic name preserves the Iron Age name of the Dunkellin river and Rahasane Lake as *Dur Abha and Dur Loch* of Ptolemy's record. Elderly people along this road to Oran Beg proudly proclaim it as the oldest road in Ireland.

It may not be as old as the Slighe Dala and Escir Riada, but that does not detract much from its ancient pedigree.

The rampart set ran astride the boundaries of Toberbrackan and Lavally, Tarramud and Ballinillaun, Coolsrahra and Moyveela (Magh Mhela = sea plain). It pressed on through Moneymore East, straddling the Ballynageeha/Bushfield boundary, crossing Oran Beg into Frenchfort. From there it traversed Glennasacul, Carnmore West, the Lydacan (Dykeland)/Lissarulla/Cregboy boundary, Gortatleva (Dykeland), and Lakeview at the Clare River, returning back to its starting point straddling the Clare river bounding Claregalway, Kiniska and Rooaunmore (rampart land).

Second and third extensions of the joint Turoe/Knocknadala/Athenry/Rath Cruacha oppidum outer ward

The second and third extensions of the joint oppidum follow through townland after townland in the same manner as the first extension. Therefore, in order to avoid monotony these will be omitted. The fourth extension, however, covers some interesting areas and will be dealt with in a little more detail.

Fourth extension of the joint Turoe/Knocknadala/Athenry/Rath Cruacha oppidum outer ward

From Murroogh (Seaboard Dyke) on the N shore of Galway Bay, 2 miles E of Galway city, the fourth outer ward rampart set ran NNW to the S tip of Lough Corrib. Bringing the lower Lough Corrib into the defensive system, it ran for over 30 miles from Grange, at the tip of the E protrusion of lower Loch Corrib, 3 miles NE of Annaghdown (10 miles N of the city), across the NE tip of Clonbrock Demesne and around the N of Kilglass. It swung SSE around Cornamuclagh (Cor na Muc Chladh = Magic Boar's Dyke hill) N of the Ahascragh river which it skirted to the Suck river N of Ballinasloe. It ran SSE parallel to the Suck river straight on to the Shannon, a 12 mile stretch. Townlands on the Suck river at

Ballinasloe, Cleagh (Dyke) Beg, Cleagh More, Cleagh Garve and Rooaun (Dyke set) on the N side of the town all proclaim the passage of the dykes. It ran down along the Shannon to Lough Derg E of Portumna, a stretch of 15 miles.

From Cloondadoobh (Field of the 2 ramparts) at Cregg Pt. on Loch Derg it bounded W across Aughta Mountain to Woodford, Lough Atorik and Slieveanore. From Corlea Bridge it hugged the Bleach river to the N tip of Lough Graney, 17 miles from Lough Derg (Figure 5.4). It swung NW through Ballardiggan (Dyke Tld) at Lough Cutra. It ran another 18 miles via Kilbeacanty and Kiltartan to Mulroog and the sea at Killeenaran near the mouth of Ath Claidh Magh Ri, opposite Muroogh in Galway Bay. Mulroog (hill of the rout) above Dunbulcaun Bay corroborates legendary history's telling of numerous incursions of hostile invaders here. This vast oppidum incorporated the sea inlets of Galway Bay within its defensive system. Numerous strongforts along this coast underpin its vulnerabity.

Figure 5.4. Lough Greine in County Clare. Vestiges of linear embankments can be found around the lake.

The most apt reference to any seaport in archaic Irish records claims: "*Ath Claidh Magh Ri: gur ab e an cuan so, **aon chuan** gabala na hEirionn mar do frioth a n-ar leabhraibh, oir is ann is mionca do hionnsuighe Eire riamh*" ("Ath Claidh Magh Ri was **the one Irish seaport** visited by sea-going vessels **through which Ireland was most often invaded**"). The number of invasions related in legendary history which came through *Ath Claidh Magh Ri* (near Clarenbridge), the great sea port of Iron Age Ireland at the centre of Galway Bay, underline the vulnerability of this massive oppidum from the seaward side. The name of the port is itself significant.

Early Dindsenchas mention of a dyke defensive system at Ath Claidh Magh Ri of Clarenbridge

The Dindsenchas of *Ath Claidh Medraige (Magh Ri)* tells of a raid by the Clanna Dedaid from Munster following the *Tain Bo Dartaid*. It speaks of Ath Claidh Magh Ri (Clarenbridge, SW of Athenry) as a place of "numerous defensive dykes (*claidhbha is coibhden*, not *cliath-atha*) and warriors" where "the Clanna Dedaid waged red strife on a foray against the 7 Maines with 3000 raided cattle. Bloodstained were the braves on raids with the kingly Maines." The Maines (sons of Queen Medb) had erected earthen black-thorn and red-thorn-covered dykes (*claidhbha draigin is derg-sciatha*) as a defence against hostile invasion. "From these defences Ath Claidh Magh Ri with its dense population is named: *'O na claidbha-sin* (not cliathaib-sin) *amne is Ath Claidh co tresse trebh.'* Since the Irish *claidh* (dyke) is generally rendered as 'Clee' orally, and sounds like *cliath* (sword), one can see how confusion crept in when scribes transmitting the oral dindsenchas version heard *'Clee'* but wrote *'cliath'*. Some authors did translate it as a thornbush-covered defensive structure. But others rendered it as thorn-bush covered spears!! Common sense demands that the original word was *claidh (clee)*, not *cliath* (sword). The same confusion reigns in the inept anglicizations of the stories of Lughaid Mac Con in his Iron Age invasion through Ath Claidh Magh Ri as told in Cath Magh Muc

Dhruime. Written descriptions, adopting *cliath* (sword) instead of *claidh (clee; dyke)*, were deemed 'obscure' and 'misunderstood' by editors and interpolators alike as was the name of the place where Lughaid's army landed, Ath Cliath Magh Ri. The outer-ward '*Claidh*' (Dyke) has been rendered correctly as a thornbush-covered defensive dyke even by editors who adopted the Old Irish *cliath* (Modern Irish *claiomh*) but badly confounded by others who attempted a cleith/cliatha/claiomha (sword) interpretation. Henceforth, the Iron Age port of Ath Cliath Magh Ri will therefore be given its correct original Ath Claidh (Clee) Magh Ri title, corresponding to Rath Clyde in Scotland.

The great stone forts on the Aran Islands played a vital role here as did the ancient forests. 'Muck' townlands, such as Muckinish (Boar Island), Muckcoort (pass of the wild boar) and Muckrush (boar forest) along this Galway coast and shores of Lough Corrib fronting this oppidum, are a reminder that the entire Galway Bay area was forest land in the Iron Age. Ossianic tales tell of Finn Mac Cool "hunting the forested harbours of Galway Bay." The ancient forest of Ochoill Oichne (Eochuil or Achaill = Yew forest), later Coill Ua bhFhriachrach, in this coastal area E and S of Galway Bay, survived into late medieval times. Forests were the greatest defensive shield and were heavily incorporated into the defensive network of this oppidum in typical Belgic (Fir Belg) fashion.

Further extensions of the joint Turoe/Knocknadala/Athenry/ Rath Cruachu oppidum outer ward

Having followed the extensions of outer wards and their relationship to one another up to this point, one could now almost logically predict the route of the next outer ward extension through the countryside. Taking into account the fact that the set of ramparts of each new extension of the outer ward has become more elaborate and complicated, one could now predict a span of approximately a mile and a half from the centre of one outer ward line of fortifications to the centre of the next.

By this time the long list of placenames, increasing excessively with every extension of the outer ward, has become such that one is in danger of getting confused with the deluge of townland names. There are many further extensions of the outer wards, with further extensions northward. These are detailed more completely in 'Hand of History – Burden of Pseudohistory'.

In the townlands through which all of these extensions pass there are repeated references to embankments in the names, as we have seen in the previous extensions. Delve in anywhere along the defence lines and be amazed by the rich oral local toponomy whose profuse names of hills, fields, forts, rivers, bogs and townlands, not only those recorded on O.S. maps but especially those enduring on local lips, proclaim the passage of the defensive dykes in many and varied ways. Take Derrybrien along the S flank for instance; local names exist there such as Cnoc na doo (Dyke hill), Clee Sceagh and Clee Thorn (Thornbush Dykes) and Loch a' Chlaidh (Dyke Lake). Others are Ming an dooan, Ming a' dooroge (dooan/dooroge = defence rampart set), Curraugh, Currach, Garrai na Clais, Claidh na Cille, Roo, Cul Roo, Cnoc na nDhoo, all Old Irish terms referring to the linear embankments. This holds true at almost any point along defence lines where old folk with long memories hand on unrecorded oral toponomy. This should be recorded by the younger generation before it is gone forever. This grass roots history, like rampart vestiges, is a history more authentically passed on through the Irish blood stream and oral memory than in the so-called history books. Time and time again the significance of geographical features and historical events appears in the townland names. The following photographs (Figure 5.5 - 5.15) show examples of some of the ramparts from various parts of the county. A map showing the approximate position of some of these embankments is shown on page 10.

Figure 5.5. Many embankment vestiges, like this one near Derrydonnell, appear as no more that a furze covered bank. These can stretch for miles across the countryside.

Figure 5.6. Avenue-like ramparts, like this, are a common feature of the Galway countryside.

Figure 5.7. This embankment is in Newgrove, Kilreekil.

Figure 5.8. This example of rampart is from the Kilconnell area.

Figure 5.9. Decayed vestiges of rampart overlooking Loughrea Lake.

Figure 5.10. Ramparts from the Doughcloon area, near New Inn.

Figure 5.11. Many lakes, like that shown, were incorporated into the defensive system. Ramparts can be found leading to the lakeshores and continuing from the opposite side.

Figure 5.12. Like that shown, many ramparts were levelled in preparation for road or rail building projects that were never completed. In this example, the main and outer lip ramparts are clearly discernible.

Figure 5.13. Rampart vestiges from the Woodlawn area.

Figure 5.14. This stretch of rampart can be seen in the Kiltullagh area.

Figure 5.15. The outer ward extensions of the joint oppiudum incorporated parts of counties Mayo, Roscommon and Clare. This rampart is in County Mayo.

Chapter 6.

Background to the establishment of Royal Centres of Power at Turoe and Athenry

The preceding chapters have presented the very real evidence of linear fortifications and forts in an area of east Galway that is full of Royal connotations in its townland names, placenames, fort names and oral history. So who were the people who constructed these great fortifications ramparts and forts? To get an understanding of these people and what inspired them it is necessary to go back to the invasion of Europe by the Romans in the Centuries B.C. and the Roman invasion of Britain by Caesar in 55 and 54 B.C.

The background in Europe

The Celtic people of Europe consisted of a host of tribes who had similar language, traditions and customs. These tribes included the Volcae, Brigantes, Atrabates etc. They originated in various parts of Europe and over centuries spread to different places. For example, the Volcae originated in the upper Danube and spread from there so that by 100 B.C. they were in the Belgium/Holland area, where they were called the Belgae. During this time they lived in relative harmony, although war was always waged against other tribes in an attempt to gain additional territory.

But all that changed with the advance of the Romans up through Europe. In 58 B.C. Caesar, Pompey and Crassus, the First Roman Triumvirate, became virtual dictators of Rome. Caesar wanted to achieve a military reputation greater than that of Pompey and was aware that the conquest of new lands would enhance his reputation. He set his eyes on Gaul (now France), the mainly unchartered lands of the Celts. Some of the Celtic tribes would have been supportive of Roman intervention as they were under pressure from expanding Germanic tribes and thought that Rome would protect them. Little did they know that what Caesar had in mind was total annihilation and domination. In the south of Gaul (France) Caesar did not have much trouble in gaining the control of the Celtic tribes. An element of surprise, coupled with a policy of 'divide and conquer', Roman bribery and highly-trained, battle-hardened, disciplined Roman legions ensured easy victory. Seeing the defeat of the southern tribes, the northern tribes (The Belgic federation of Celtic tribes), which had formed a federation where they joined forces against a common Germanic enemy, developed a new defensive strategy. They had learned valuable lessons from the failed military strategies of their southern Celtic cousins and by the years of border conflict against Germanic tribes. Their territory was bounded on the S by the Marne and Seine, on the W by the sea, on the E and N by the Rhine. This provided no protection as deep incursions by Germans had shown. They had seen how hillforts were turned into death traps by Caesar's tactics of surrounding them. New military strategies and defensive systems were necessary to outwit Rome's formidable force. They could never defeat highly-trained, battle-hardened, disciplined Roman legions in set battle. New methods of guerrilla warfare were devised by employing a radically new system of defences, incorporating forests and marshes into a web of complicated earthworks into which the Roman troops were lured, unaware of lurking dangers. The more southerly Belgic tribes had no time to implement this new defensive system before Caesar suddenly attacked them. The more northerly Belgic tribes (particularly the Menapii) on the Continental Lowlands held out due to their radically new defensive tactics, which took Caesar and the Romans by complete surprise. Others like the Atrebates and Morini made tactical temporary submission to gain time to complete this system. However, in the long run with the connivance and assistance of Germanic tribes the Romans eventually overcame the Belgic Celts, many of whom fled to Britain, which was a short journey across the North Sea.

The emergance of Commius as a Celtic leader

During this Roman invasion of Europe, Commius, a Celtic leader, came to prominence. His influence over the Celtic tribes in Europe, Britain and possibly Ireland was enormous and he made a considerable contribution to the defence policies.

Bituitus, King of the Arverni (a Celtic tribe), was imprisoned in Rome for failing to submit to Roman authority. After years in Roman custody, Bituitos' son, Congentiatos, was born and kept in custody with his father to ensure he would not claim the Arvernian throne, now abolished by Roman order. The name Congentiatos (born of a divine dog of war) is reminiscent of the name of his grandfather, Louernius (wolf or fox god), a typical Celtic name for a born fighter. Congentiatos proved too long-winded a name for general use and he became known by the more homely 'Conngenn', 'Gann' for short. The Romans did not like the Celtic connotations of this name and called him 'Commius' instead. It remained his official name in Roman circles. The date of the imprisonment of Bituitus and of the birth of Commius (Congentiatus) are unknown.

Commius was given a classical education which he would not have had were his father not a prisoner of Rome. Having completed his classical Roman education, he joined his kinsmen in exile rather than remain in Rome. He fled N and found asylum among the Belgae, as his father had done before him. Commius' sterling leadership qualities were highly prized. He had a charismatic personality, which commanded Caesar's respect. When Caesar overcame the resistance of the Belgic Atrebates by expelling their anti-Roman chiefs, he appointed Commius to be their king. "Commius" he said "was a man both courageous and politic,...a noble of great influence...and learning." He became Caesar's 'right hand man', but at the same time acted as a double agent for the Belgae as his tribal affinity was hidden from Caesar. Importantly, Caesar used Commius as negotiator in preparing for his invasion of Britain and as intermediary in discussions with the Britons. Through him Caesar would receive the submission of Cassivellaunus whom the British Celts had chosen to command their combined forces. For this service Caesar freed Commius from tribute, restored the rights and laws of the Atrebates lost by their earlier resistence to Rome. Because of his influence, nobility, learning and leadership qualities Caesar gave him sovereignty also over the Morini, a neighbouring state on the Belgic sea-board. Commius had the astute ability to have himself accepted as a highly respected leader of diverse Celtic tribes. He was an outsider uniquely acceptable to all. Later he was appointed joint Overking not only of Continental Belgic tribes but over their kinsmen, the Belgic Atrebates and Morini in SE Britain.

The Roman invasion of Britain

Caesar's plan to invade Britain was formed in 57 B.C. when fighting the Belgae. He found that several Belgic clans had fled to SE Britain, or were about to do so, rather than submit to him. He noted that early British Belgae migrated as conscious offshoots of continental tribes and retained the names of the Continental tribes from which they sprung and maintained lively contact with their mainland cousins, aiding them in the war against Caesar. He used this fact as an excuse for his invasion of Britain. Caesar distinguished the Britons of the maritime parts, who preserved a tradition that their ancestors had crossed from Belgium (*'ex Belgio'*) from those of the interior, who had been in the country 'from time immemorial' (a few generations earlier). In Britain, the Belgae drove the aboriginal Priteni and early Celts north. In 57 B.C. the Belgic Bellovaci fled to the Belgic area of Britain with their followers. Belgic tribes, such as the Atrebates, settled in Britain, before and after Caesar's campaigns in 55/54 B.C.

On 'expert' advice Caesar chose 10,000 men for a 'reconnaissance' trip to Britain. With such an unusually large force to be transported by sea, it looked as if Caesar had hoped this would be his conquest of Britain. This was his last chance for a full-scale invasion before his term of office as Governer of Gaul was up. His record of the event is an excuse for his failure, making it appear that it was never meant to be more than a preliminary 'reconnaissance'. Some 80 transport vessels were assembled and set sail. Commius was sent ahead to negotiate a submission from the British Atrebates and other Belgae because of his influence over them. But he turned tide and warned the British Celts of the imminent danger. Thousands of Celtic warriors massed all along the cliff-tops and when Caesar eventually landed, battle ensued. One Celtic weapon unnerved the Romans, encountering it for the first time: the expertly handled war-chariot, with scythes fitted to wheel hubs it mowed down the enemy. Warriors raced those deadly chariots through the shallows, petrifying the invaders, then engaging them in hand to hand combat before darting back on to the chariots and racing away. Caesar admitted in his records that his men were dismayed and disorganised by this unaccustomed manner of fighting.

Caesar was in trouble and after a short while chieftains of the Cantii (a Celtic tribe in Kent) arrived at Caesar's camp, with Commius allegedly as their 'hostage', to open negotiations for a truce. Commius had warned them of how easily Gaul had crumbled before Caesar. He told Caesar he had been 'imprisoned' by Cantii chieftains to prevent his warning the Romans that the Celts were ready to fight Caesar on the beaches. With great difficulty Caesar's much depleted army, low in morale, hungry and without provisions, limped back to the Continent. From a Roman military viewpoint the expedition was a disaster. Yet, Caesar did have the opportunity to see for himself just how expert the British Celts were at using their war-chariots, an experience he did not encounter on the Continent. He gained valuable, though costly, experience, but had seen nothing of the land. He insisted it was meant to be no more than 'a military reconnaissance' in preparation for full-scale invasion. One hard lesson learnt was that 10,000 chosen legionaries was insufficient even for a mild reconnaissance. As he was retreating from the British coastline Caesar was already planning a 'full- scale invasion' and 'conquest' of the entire island with a sufficient force which he now realized would have to be very much larger than that of his first attempt. He convinced Rome that the conquest of Britain was all but a reality. The Senate voted him a further five years on the strength of his successful report of his first invasion attempt. During the winter of 55/54 B.C. Caesar prepared his for the conquest of Britain in the coming year.

In April 54 B.C. Caesar had 800 transport ships, 28 warships and 30,000 men ready for the invasion. This kind of preparation could not be kept secret and rumours of the forthcoming massive invasion spread quickly. In contrast to the previous year, the landings were unopposed. "The enemy was nowhere to be seen" Caesar wrote. Captured prisoners informed him that initially a large army had gathered to meet the Romans on the beaches, but on observing the vast size of the Roman fleet, they retired inland.

By the 7th of July the Roman army had established themselves on shore. They marched inland and came in sight of the Celtic forces who had established a strong position on the banks of the Great Stour river. This was the land of the Belgic Cantii. From high ground above, Celtic war-chariots came racing down harrying the Romans as they attempted to cross the river. With fire from Roman archers and bowmen they forced the Celts to retreat. Caesar noted that they then retired to the woods where they had a fortified position of great natural strength. This was Caesar's first experience of the British Belgic oppidum defence system. Some Celts retreated into the nearby fortification surrounded by forest and other defensive features, inviting the Romans to follow, while others attacked the legion's passage through wooded areas. The Celts had prepared an ambush by having most of the approaches to the fortress obstructed by the felling of trees. Small groups of Celts harried the Roman advance all along the route.

Caesar inspected the Belgic oppidum fortifications. Palisade-topped ramparts rose just 20 feet above the base of the wide deep defensive ditch surrounding the earthen embankment. He ordered a legion to storm the hillfort as bowmen and slingmen started firing to keep the heads of Celtic defenders down while a Romans piled up earth against the embankments in order to scale them. When complete, the Romans poured over the rampart into the fortress only to find a deserted fort held by a handfull of disappearing defenders. This was of the essence of Belgic defensive tactics. Far from being defeated, Cantii chiefs and their main forces lay hiding in the woods waiting to engage the Roman legions in guerilla warfare.

This type of warfare slowed the Roman progress inland, although they were eventually victorious over the Cantii. As they marched inland they realised that the Belgae and Catuvellauni (who were 70 miles from the sea) had erected what was for Caesar and his legions an entirely new type of defensive system whose primary purpose was aimed at guerilla type warfare. The Celts would feign retreat drawing the Roman forces into unfamiliar dense woodlands or marshy terrain. There they were easily separated out of view of one another and of their commanders, sink in quagmire under the weight of heavy encumberances, and where they could no longer fight in tightly-knit formation as was their custom. Caesar surprised the southern Belgic tribes of northern France and Belgium before they had time to exploit this new military strategy. His military strategy was one of lightning-fast mobility to catch Celts off guard, unprepared, with no choice but to surrender or perish. The new Belgic defensive strategy, adopted by the Belgae and Catuvellauni brought large tracts of forest and marshland within the fortifications into which the enemy might be lured, ambushed, harried, worn down and delayed, using every conceivable guerilla tactic. Thus, there was a shift of emphasis in strategic control of the region from the smaller hill-forts to the more extensive Belgic defensive system of the sprawling oppida settlements.

All Celtic tribes of SE Britain had unanimously agreed to confer the supreme command of operations against Caesar upon Cassivellaunus, the Catuvellaunian chieftain. His capital, the central stronghold of his oppidum, was the massive hillfort at Wheathampstead, a little north of St. Albans. His primary objective was to keep his forces steadily in retreat before the Romans, thereby drawing them within target areas of his extremely complicated system of fortifications where they would be entirely at the mercy of his troops hiding in ambush in the woods, operating in small bands.

Caesar painted a glowing picture of the strategy and tactics being implemented by the Belgae and waxed lyrical on the effectiveness and sophisticated thinking behind the uniquely new Belgic defensive system incorporating forests, marshes and rivers over a wide stretch of territory: "In the whole of this kind of battle, since it was fought under the eyes of all in front of the camp, it was perceived that our men, on account of the weight of their arms (in as much as they could neither follow those who were giving war, nor dared to depart from their standards) were little suited for an enemy of this kind; that the cavalry manoeuvre fought with great danger, because the enemy would oft-times retreat even designedly, and, when they had drawn off our cavalry a little way from the legions, would leap from their chariots and fight on foot in unequal combat...The enemy never fought in close order, but in small parties and at considerable distances, and had detachments placed about (like substitutes on the sideline), which, in turn, took the place of others, so that vigorous and fresh troops continually succeeded those who were weary" (*Caesar's De Bello Gallico*). The brilliant hit-and-run guerilla tactics employed by the Celts took Caesar totally by surprise and took an enormous toll on his massive army. Thanks to the complicated Belgic defensive strategy!

Eventually, the Romans crossed the Thames and captured the Catuvallauian stronghold. But it took its toll on them. Caesar soon realized that while Cassivellaunus refused to submit, the war was not won. There were a dozen similar strongholds throughout Britain. Caesar did not relish the prospect of having to reduce them one by one. Moreover, Cassivelaunus had not been defeated and still exerted farflung authority. By the time Wheathamstead was captured it was August and it was time for talks of a truce. But it was Cassivelaunus who instigated the peace talks. Caesar wrote with satisfaction: "So many losses having been received, his territories devastated, and being distressed most of all by the defections of the tribes, he sent ambassadors to me through Commius, the Atrebatan king, concerning surrender."

A major factor behind the withdrawal of Cassivelaunus was that the Trinovantes, whose tribal territory bordered on that of the Catuvelauni to the east, accepted a pro-Roman as their king. Caesar was now welcome in their territory. They in turn provided provisions, men and logistic support for his army. In return Caesar treated them honourably as Roman citizens. Seeing this, other chieftains and tribal warriors decided that surrender was the best way to preserve themselves and their territories. They withdrew their forces and returned to their tribal lands. This was a severe blow to Cassivellaunus, who was now unable to prevent the Romans from obtaining supplies.

Commius again emerged as a mediator. Caesar was clearly relieved to be in a position to dictate his demands. These were brief: "I demanded hostages, I demanded fixed annual tribute payable by Britain into the Roman treasury, and I strictly forbade Cassivelaunus to interfere with the Trinovantes. After receiving their hostages we returned to the coast." It was late September before Caesar and his much depleted army returned to the Continent. After 3 traumatic months in Britain, Caesar's report to the Senate of his alleged victory conceals conciousness of failure. In Rome there was public dissappointment at a victory apparently thrown away. Caesar's invasion of Britain was a financial disaster. The moral victory went to Cassivelaunns and the British Celts.

Caesar's ambitions for annexation of Britain did not end there. He was determined to return for a further campaign the following year. However, on his return to the continent the situation in Gaul drove all such thoughts from his mind. In 53 B.C., a massive Celtic revolt erupted in Europe. Commius sided with his fellow-country men against Caesar. He was elected one of the 4 supreme chiefs committed to take charge of the war. This shows his considerable standing and influence, not just among the Belgae, but right down through all Gaul where he was highly respected, most notably by Vercingetorix, Arvernian king and overall leader of the revolt. In the dramatic suppression of the revolt, Commius was the sole leader to refuse to submit, calling the Germans to his aid. When all failed he took refuge among the Germans. He was the most wanted man in Europe with an enormous price on his head. Rome failed to apprehend him. After many miraculous escapes he secretly slipped through Belgic lands, inciting the Belgae to join him in a new Belgic exodus to Britain. There he was destined to reign in peace for 20 years.

This uprising against Rome cancelled any plans Caesar might have had to invade Britain again. Britain was to retain its independence for a further century. Cassivelaunus returned to his position of pre-eminence among the British chieftains. He took advantage of the situation in Gaul to ignore the provisions of the treaty with Caesar. No annual tribute was paid to Rome. No further hostages were sent. Britain returned to being one of the most prosperous of the surviving independent Celtic countries. The subsequent history of the Catuvellauni shows that Cassivelaunus and his comrades in arms were the real victors over Caesar.

The Belgic 'invasion' of Ireland

On persistent rumours of further Roman invasions of Britain, Commius handed power over to his son, Tincommius. His last recorded act before abruptly disappearing from the British scene and Roman records appears to have been the rallying of sons and kinsmen to lead Belgic noncombatants out of harm's way, concerned for their safety. The exit of Commius from Brito/ Romano scene and records, together with his second son Dela, coincided with the exit of the Morini. It was also then that the Fir Belg invasion of Ireland took place. It may have been then that Commius/Gann led his Belgic Atrebaten (Athreibh) and Morini (Mac Morna) subjects to Ireland? The 1st/2nd century Greek Geographer Ptolemy placed the Gangani (Gann's descendants) around the mouth of the Shannon precisely where Irish legend placed the Fir Belg invasion of Gann. This Belgic incursion evolved from a safe haven into a full-blooded invasion northwards into Connaught. Connaught Belgic tribes claimed descent from Gann and Sengann well into historic times. Greek, Roman and Irish records fuse.

There is no absolute proof that the Commius elevated as King of the Continental Atrebates and Morini by Caesar was Commius, son of Bituitus, former King of the Arverni, or Commius who emerged later as King of the British Atrebates, or Gann who led the Fir Belg to Ireland. Yet, this is no mere conjecture. Several factors support this identification, not least Celtic and Roman forms of his name. When finally he became Caesar's bitter enemy he renounced his Romanised name and reverted to his original Celtic name, Congenn, Gann for short. This affinity of names impells one to identify the Arvernian Commius with the Commius, King of the Atrebates and Morini, and the Gann who led these Belgic 'Gangani' to the West of Ireland. It is further corroborated by Ptolemy's Irish Record compelling one to accept that it was Commius and son Tincommius who, when they abruptly dissappeared without trace from the British scene and Roman records due to rumors of Roman invasion, emerged in Ireland as leaders of the Belgic invasion and finally appeared in their true Celtic colours as Gann and Sengann. Legendary history names Gann and Sengann as leaders of the Fir Belg invasion of SW Ireland precisely where Ptolemy's record located the Gangani, Gann's subjects. The identification of the Irish Gann/Sengann with British Commius/Tincommius can hardly be labeled pure coincidence or conjecture. When Irish legendary history is thus ever so subtly corroborated by Brito/Romano history, Greek geo-political records (Ptolemy) and the ominous contemporary Celto/Roman military menace in Britain, then legend ceases and history begins.

Commius was then an elder some 60 years on. His son, Tincommius, and grandson, Verica, succeeded each other on their father's throne among the Atrebates in SE Britain. Both disappeared mysteriously from the pages of Brito/Romano history, like their father and grandfather. Under Celtic forms of their names, Sengann and Ferach, both re-emerged in Irish legendary history as Fir Belg kings, succeeding their brother/uncle Dela, 1st Fir Belg Overking, according to Irish tradition, to set up his throne on Knocknadala/Turoe (Ptolemy's REGIA e TE[*mh*]RA) in Co. Galway in western Ireland. Connaught tribes, such as the Partrige, claimed descent from Gann and Sengann well into historic times.

As these Celtic leaders moved to Ireland, they feared a Roman invasion of their new home. Through their experiences in northern Europe and Britain they had seen the advantages of the Belgic defensive mechanism and also witnessed some of its failings. In their new land (south-west Ireland) they were going to be prepared for a Roman invasion and set about building an extensive fortification system that was even more complex than that in Britain. This had the advantage that they were strategically more advanced than the native Cruthin and were thus able to capture their territory and protect it.

It is the remains of this Belgic defensive system that still scars the countryside around Galway. The features are unmistakable. Large double ditches sometimes in sets of three, thorn-bush lined ramparts, raised river banks to flood extensive areas and dykes built to bring bog, marsh, lake and forest into the defensive system. The size and complexity of this defensive system is unmatched anywhere in Europe, although at present it remains largely unrecognised.

Chapter 7.

Evidence for the existence of a Royal Centre of Power in County Galway

Before the coming of the Celts, particularly the Belgae (Fir Belg), Ireland was occupied by the Cruthin who became known as the 'Ulaid' as they were driven N until only Ulster remained in their hands. England was occupied by Priteni, who were driven N into Scotland from throughout the whole of England and became known as the Picts. They gave their name to the Prittanic or Brittanic Islands, which were known to the Greeks in the 5th century B.C. As we have seen in Chapter 6, there is ample evidence that Celtic tribes fled the invasion of Britain by the Romans. Since Europe was also under Roman control, Ireland was the most probable destination. While there is no direct evidence linking the British Celts to Ireland, there is an abundance of circumstantial evidence, as shown in the first 5 Chapters, leading to the West of Ireland as the main Centre of Power of the Fir Belg. This Chapter will draw together the strands of this evidence and show that far from being a 'backwater' of Iron Age Ireland, the West was a driving force in Iron Age Ireland.

Archaeological evidence

The previous Chapters have detailed the archaeological evidence from Turoe/Knocknadala and Athenry and shown the existence of an extensive series of linear fortifications in the area. As described, it is likely that these were defensive embankments and that they were associated with other features in the area e.g. roads, ringforts etc. See additional examples of the linear embankments from the area in Figures 7.1 - 7.10

Figure 7.1. A section of rampart in Cross townland.

Figure 7.2. A long transverse rampart on the south of Turoe townland.

Figure 7.3. Remnants of rampart on the north slopes of Turoe Hill.

Figure 7.4. Many rivers, like this one in Lackafinna South, had raised banks to assist in flooding large areas of low lying land. These raised banks are clearly shown on the 1933 O.S. map (see insert). Map courtesy of Ordnance Survey Ireland, Permit number MP 0001903.

Figure 7.5. Many ramparts, such as this one in Brackloon, were levelled for roads that were never completed.

Figure 7.6. Section of rampart in Kiltullagh.

Figure 7.7. A thornbush-covered rampart across Greyford townland, near Kiltullagh.

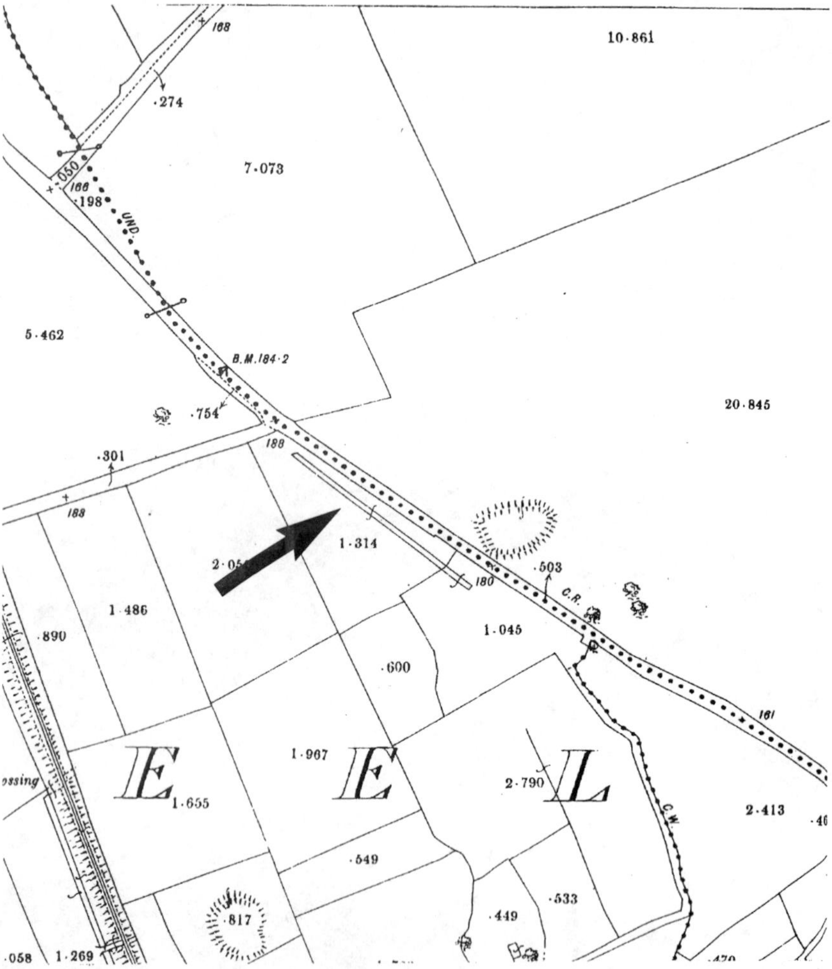

Figure 7.8. A section of rampart, now almost totally destroyed, at Knocknadala (near Forge Hill). Map courtesy of Trinity College Dublin.

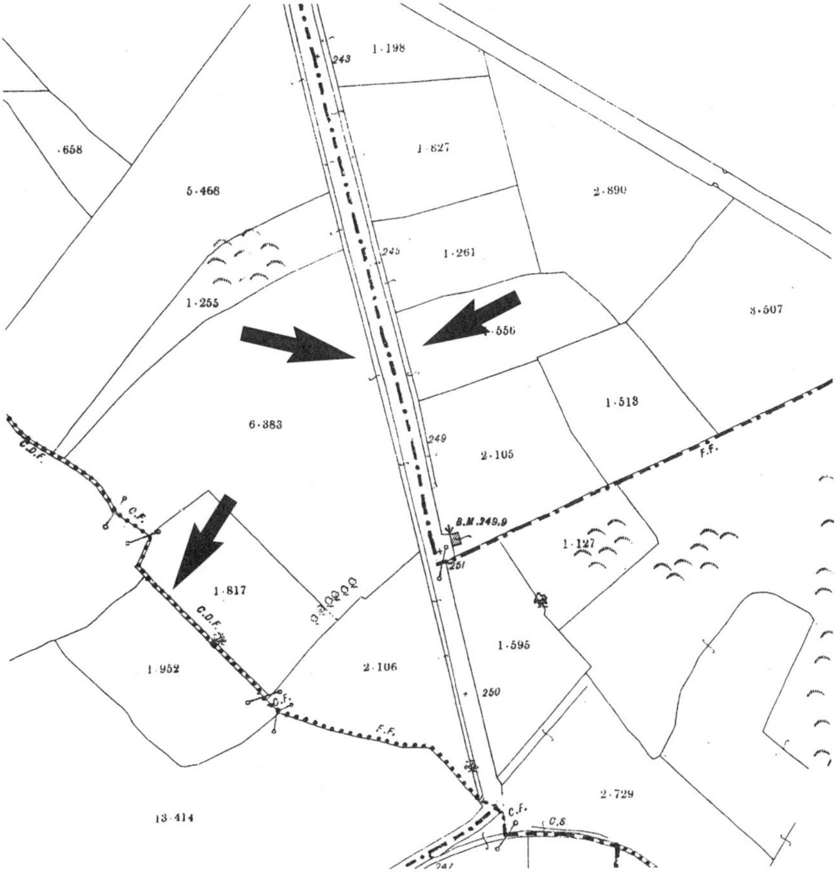

Figure 7.9. This 1894 map shows the embankments that still exist inside this road just outside Bullaun village on the road toward New Inn. The townland boundary on the left is another large embankment that continues for a long distance. Map courtesy of Trinity College Dublin.

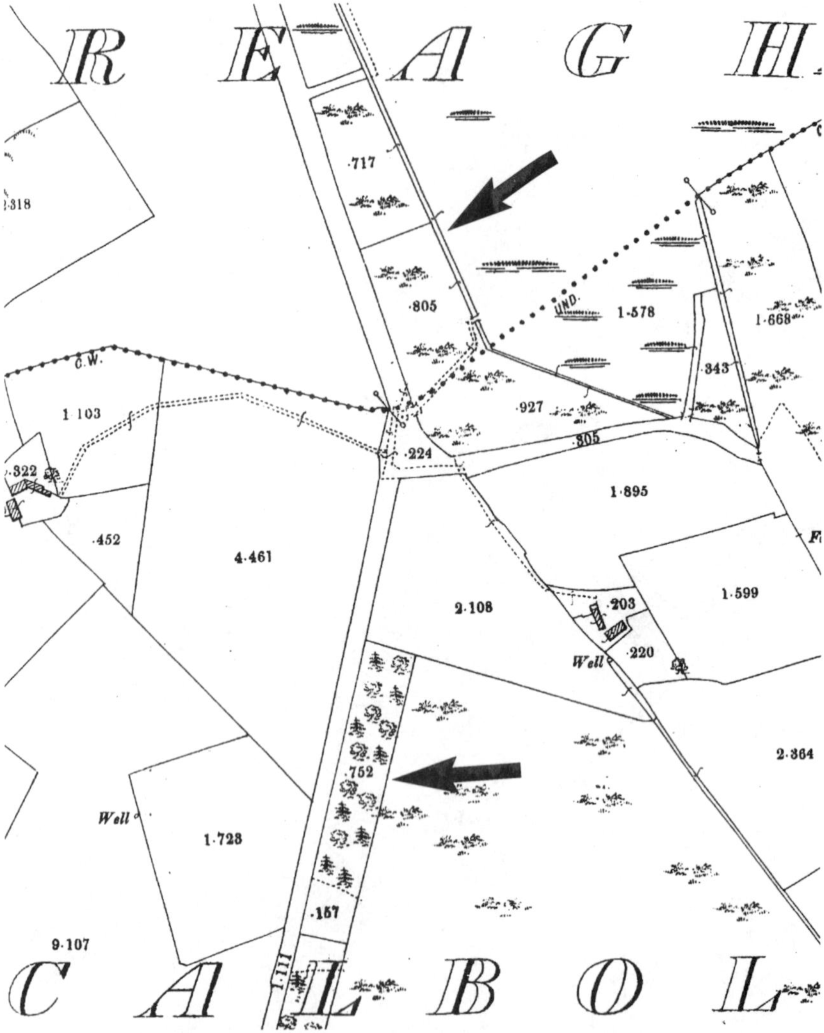

Figure 7.10. Embankments at Carrowreagh and Galboley, as shown on the 1894 O.S. map. Map courtesy of Trinity College Dublin.

Irelands early road system

The 5 great roads of ancient Ireland were said to have radiated out from Tara across the face of Ireland. When 19th Century antiquarians were called on to reconstruct and name the ancient sites of Tara the ancient highways were recorded on the restored map of Tara sites. These roads were not recorded on the Ordnance Survey because no trace of them could be found on the ground. From actual evidence on the ground all one could say is that a track, possibly ancient, leads from near the N end of the Banqueting Hall towards Rath Grainne. This is the extent of the evidence for the 5 great roads of Ireland at Tara, nothing on any map, no local traditions and no archaeological evidence. Now let us take a look at the ancient road system around Turoe. There is a complete ancient road system surrounding Turoe/Knocknadala that exists on O.S. maps, in the traditions of the area and on the ground.

One of these roads is Escir Riada (Figure 7.11). Near Esker (Athenry) there is a stretch of ancient road knows as Escir Riada. This is remembered locally as a much longer stretch and was recently partially demolished. This old road headed East through Carnakelly, Gortakareen and Raford. At Raford it was called the Sean Bhealach. It continued towards Knocknadala, where a short stretch still remains a sunken roadway. From Knocknadala it headed East along the sand hills. This sand hill which has recently been demolished in gravel mining was known locally as Escir Riada. (See Chapter 2)

Intriguingly, legendary history claims that Slighe Dala ran from Temhair na Ri (could be interpreted as Turoe, since 'Temhair' is Irish for Turoe and Tara) to Roscrea and Borris-in-Ossory. The same legendary history claims that this renowned highway was founded by Dela, alleged father of the leaders of the Fir Belg invasion. The same legend further related that Roscrea, its original S destination, got its name from Crea, the Munster Fir Belg Princess who became Dela's consort.

Figure 7.11. As shown by this section near Caherdrine ringfort, the Escir Riada ran in a straight line for long distances. In this section, the rounded banks have been replaced with stone walls.

There are local traditions of the Slighe Dala around Turoe and along its route from there to Roscrea. It ran south across the hills of Knocknadala and Turoe to Loughrea. A short stretch still survives, albeit in a dilapidated state, just N of Ballykeeran Bridge (Factory Bridge). The present road along Turoe's W boundary with Ballykeeran is built on the route of this ancient highway. In Ballykeeran townland remnants of its low wide banks there were levelled by locals fifty odd years ago. From the SW corner of Turoe, Slighe Dala continued straight on S to Loughrea, leaving short sections of this ancient highway with banks intact alongside Tonnawausa fort in Lackafinna, and again in the field across the road from the Rugby field near Loughrea. At Loughrea it swung E on the heights above Loughrea lake to the massive ringforts of Masonbrook. A long section survives here. Then it swung ESE past the Hill of Doon on its way to Portumna and Roscrea. Local elders along the course of the Slighe Dala, some on their last legs, helped locate the route of this revered ancient highway. Some could indicate only short stretches. A rare few like Joe Cooney, Michael Gleeson, Joe Dunne, Deely, Ruane, and Casserly, were not only able to point out its route for long distances even to its terminus, but also to pass on its ancient name. It had several local names but only one overall name, Slighe Dala. Further afield near Portumna and E of the Shannon it was called Bealach Mor Muigh Dala. Other names for Sli Dala on local lips are Sean Bealach, Bealach Mor, Slighe Mor Dala, etc. Several stretches of this ancient road are still traceable confirming Mr. Cooney's alignment. This ancient highway from Knocknadala to Roscrea was still known by many old people in Mr. Cooney's young days as Sli Dala. It ran in a straight line for miles on end. Wider than most, it is 21 ft. wide at all points where it survives intact, excluding its earthen banks. The stretch from Knocknadala via Turoe to Loughrea was still used as a shortcut until early in the 20th Century when a bridge was demolished in flooding. (See Chapter 2)

'Rot na Ri' is another ancient road both on elderly local lips in the Turoe/Knocknadala vicinity and in archaic legendary material such as the poem "Magh Fot ar Rot na Ri", which corroborates the route of the road through Moyode. There are apparently two routes from Turoe/Knocknadala to Myode. One came from Cotni (Carrowkeel) and followed the now destroyed esker running W to the river near the castle in Cloghereavaun and ran W along the N bank of the straight stretch of river from Clogharevaun Bridge. It can be followed for a good distance W of the bridge leading towards the very outstanding Esker a bit further W which runs all the way through Greyford, Curraghmore and Rathgorgin to Moyode. The second route starts from Gortakeeran, ran mostly along the present Raford road to the forge at Kiltullagh, along the old narrow road behind Kiltulla Church and on along the road to Kiltulla cemetery. It continued straight on across the fields before joining the main Kiltulla/Athenry road at or near Curraghbeg and then continuing straight on through Rathgorgin to Moyode. From

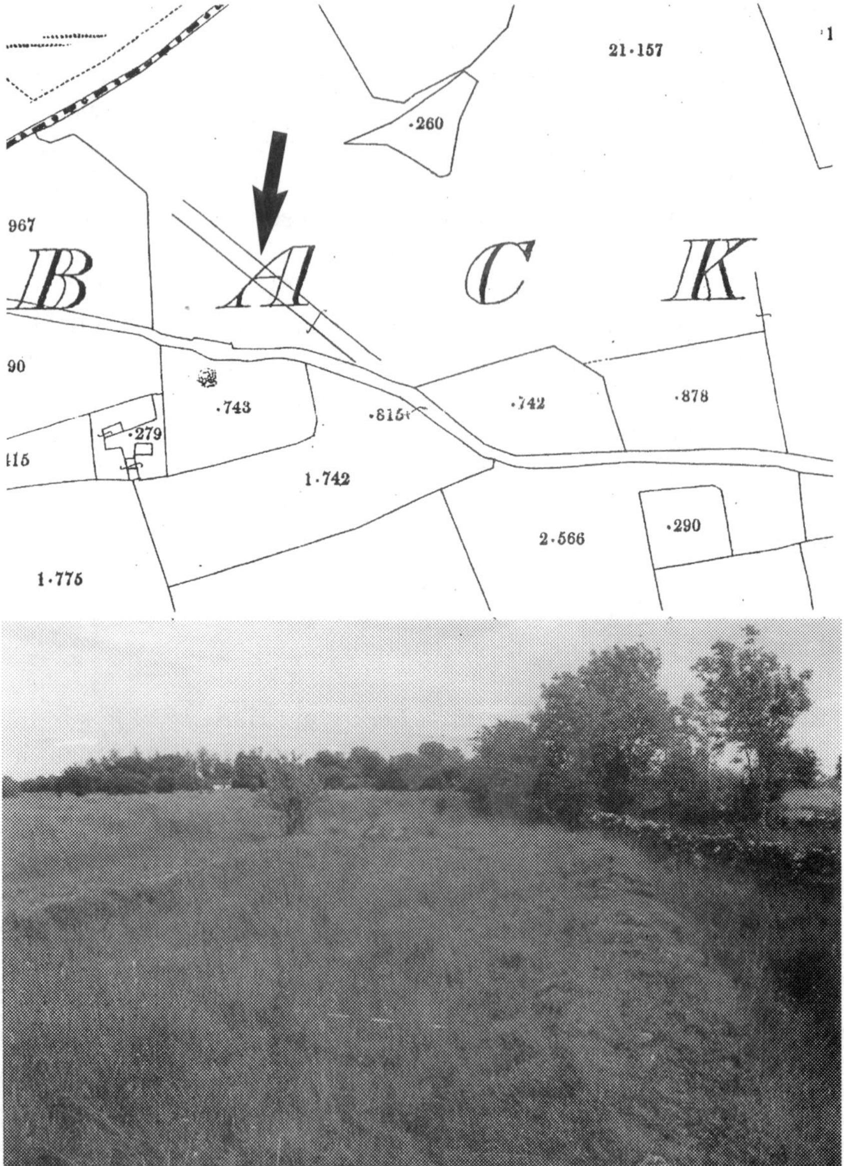

Figure 7.12. A section of ancient roadway in Backpark, near Athenry. With rounded banks, this is similar to other ancient road sections. This is shown on the 1894 O.S. map (above). Map courtesy of Trinity College Dublin.

Moyode this road passed through Ballywinna, Parkroe and on to Clarinbridge. This was a shorter route to Ath Clee Magh Ri than the Escir Riada and may have only been used by Kings, hence the name 'Rot na Ri (Road of the Kings). Many modern byroads lie along several sections of this archaic road and are notable for their straight alignment.

The final road in the Turoe tradition is Sli Coolan. In more recent times this road was known as Sarsfield's Road as it was used by Sarsfield in his retreat from Limerick. This road went from Galboley, westwards around a hill called Cruachain Coolan, passed the Ceremonial Avenue leading to the Turoe Stone and joined the Sli Dala on the border of Ballykeeran and Farta. Traces of this road exist as an avenue beside Keanes in Farta on the West and on the East as part of the Galboley road. Although there is no trace of the centre portion, its route could be pointed out by locals.

Another archaic route branched off from Sli Dala near Loughrea and ran S through Clare. Mr. Cooney claimed that Irish speakers of the older generation in his young days called it 'Sli Luachra'. The Munster terminus of this road was the Fir Belg Capital of Temhair Luachra 6 miles due S of Limerick city.

Other sections of ancient road that can not be linked to the previous roads exist, such as the section at Backpark near Athenry (Figure 7.12)

Archaeologists have failed to trace any of these great roads in the vicinity of Tara, but the traditions of ancient roads are still alive around Turoe. These traditions are corroborated by the archaeological evidence of sections of ancient roads in the area and longer sections that are now destroyed, but are shown on old O.S. maps. This calls for a thorough investigation of the ancient road system, not only around Turoe and Knocknadala and Athenry, but of Ireland in general.

Documentary evidence

Similar to the evidence of ancient roads, there is no shortage of documentary evidence relating to Turoe/Knocknadala and Athenry as being an important Royal Centre.

Ptolemy of Alexandria was a well respected cartographer of 1st/2nd century A.D. He is well known for the accuracy of his work, which has survived to this day. A map of Ireland showing the position of rivers, towns and tribes, which can be drawn from co-ordinates left by Ptolemy, has survived. The north, east and south coasts are recognised as accurate, but the west coast, because it is thought that seafarers of the day would not be familiar with it, is thought to be inaccurate. The west coast as it is described by Ptolemy does not fit the preconcieved view of Ireland at the time i.e. that of Tara as the seat of the High Kings. However, it is now known that the seafarers were most familiar with the west coast of Ireland. No other Iron Age sea port is more renowned than Ath Cliath Magh Ri. The most important reference to any seaport in archaic Irish records points directly to this western port with astonishing precision: "*Ath Cliath Magh Ri: gur ab e an cuan so, aon chuan gabala na hEirionn mar do frioth a n-ar leabhraibh, oir is ann is mionca do hionn- suighe Eire riamh*" ("Ath Cliath Magh Ri was **the one** Irish seaport visited by sea-going vessels and through which Ireland was always most often invaded"). Archaic references state that "all ships sailing the high seas called at Ath Clee Magh Ri." Surely it was to this port in particular that the Roman author Tacitus was referring when he made the astounding claim that no British port was better known to sailors than those of Ireland." The stormy Atlantic coasts of the west were, not less, but rather most familiar to Iron Age merchants. Therefore, Ptolemy's record of the Capitals of Iron Age Ireland is likely to be correct.

Ptolemy recorded two Royal Capitals (REGIA) in the Ireland of his day. He placed one of his Royal Capitals in the NE of Ireland simply as 'REGIA'. This is generally identified as Ulster's ancient Royal Capital, Emain Macha, in Armagh. He located his 2nd

'REGIA' at the centre of Co. Galway in the W of Ireland. His record in its present state has 'Regia etera' which should be amended to its original 'Regia e Tera' (Regia e Te(mh)ra). This can be interpreted as 'Royal Capital of Turoe'. It is extremely significant that neither of Ptolemy's 'Regia' was Tara of Meath. Tara, the so-called 'Royal Capital of Ireland from time immemorial', is conspicuous only by its absence from Ptolemy's record. Tara was unknown as a major Royal Capital to Ptolemy or his informants as it did not become important for centuries afterwards. Also on Ptolemy's record is a river called the 'Dur'. This is extremely significant as the Dughiortach (Dunkellin) River is also called the 'Doreythra', which is certainly related to 'Dur'. The turlough that this river flows into, while now called Rahasane Lake was known to elderly locals as 'Dur Lough'. While the 'Dur' river is in the wrong place on Ptolemy's record to equate to the Dughiortach (Dunkellin), so to is the 'Senos' in the wrong place to correspond with the Shannon. The list of names along the whole west coast of Ireland as given by Ptolemy have been read from top to bottom instead of the other way around and so the map is generally drawn upside down on the West. When read in the right manner all the names fall into their right positions, including the Shannon and the Dur (An Dughiortach). The whole west coast of Ireland given by Ptolemy has been incorrectly read from top to bottom, rather than from left to right, which accounts for the Dur and Sennos replacing one another.

Other documentary evidence comes from Irish legendary history. The 'Dindsenchas' documents are important in this regard and make several references applicable to Turoe/Knocknadala. One legend states that Dela (with 2000 Fir Belg warriors) set up his throne on Dail and established the Royal Oenach beside it: "*Adnagar Delach ar Dail. Roghni Oenach Dind re thaibh.*" The legend then states that they captured 'Temhair', held till then by the Cruthin. Not much attention was paid to this legend as it did not fit the preconcieved idea that 'Temhair' was Tara of Meath and therefore, geographically, it was not right. However, Turoe of Galway and Tara of Meath share the same Old Irish name,

'Temhair', Therefore, this legend could apply to Turoe with Conc na Dail beside it. In 'Fiannaigheachta' texts its name is given as Cnoc na nDal where it is referred to as one of the favorite places frequented by Finn Mac Cool. As we have seen, this is the only hill in Ireland with the name meaning 'Hill of Parliament' or 'Hill of Great Assemblies'. Parliament Hill Fields, N of the Thames in London, was Celtic Britian's Cnoc na Dail. As the name signifies, it was the meeting-place of the great assembly of the Celtic tribes. Here one is at the very centre of the Over-kingdom, where the tribes assembled, whence they were governed, and where their laws were enacted.

The 'Dindsenchas of Medraige' (Maree, Co Galway), gives a significant short genealogy of a grandson of Dela which ties the spearhead of the Fir Belg expansion against the Cruthin to Turoe of Galway: "Fermhor (Ferach Mhor) son of Eremon (Ercmon or Erc), son of Dela, son of Genann (Gann)." Gann was one of the Fir Belg leaders who set up around the Shannon, where the Gangani are recorded by Ptolemy.

Legends further relate that Eochaid Fedlech came north across Fremand (*'tar Fremaind adthuaidh'*) to take over the kingship of Temhair (this could apply to Turoe). It further claims that he was the first King of the Fir Belg to reside on Temhroit (*'ced rig do suid ar tus i Temhroid de Fhearaibh Bolc'*) and also last to set up his royal residence there. This Fermhor is variously called Eochaid Ferach Mhor, Eochaid Fedlech, 'Ri Temhro' (King of Temhair), Ideal King of the Fir Belg, and father of Queen Medb. There is no Rath Ferach Mhor on the hill of Tara and Tara was never known as Temhroid de Fhearaibh Bolc. So all of this could apply to Turoe as the ringfort on Turoe Hill where the Turoe Stone originally stood is called Rath Firmhor (Ferach Mhor).

The area of Cruach Boidbh (Boyhill) of Athenry is very significant. Boidbh or Badbh was a celebrated Celtic war-goddess, referred to as Badbhcath or Cathubodhua in the Continental inscription of Haute-Savoie. It is interesting that from the NE of Athenry, E of

Laragh Gate, an ancient road known locally as Bothar Bhodbh (corrupt 'Bothar Vov' or B. Woh') followed the route towards the present Lady's Well/Rahard road and straight up to Boy Hill (Cnoc Buidbh). Sections of Bothar Bhoibh survive, adjacent to Boy Hill (Figure 7.13). According to Aggie Qualter, it was a popular and lovely walk for townspeople early in the (20th) century. It was bounded by beautiful rounded banks. Stone walls replaced these and narrowed its original width. It now stands forlorn and forgotten.

The significance of this area is even greater as the Book of Leinster relates that Ochall, Cruachan and Sidh Buidbh were close together. Further text also refers to 'Relig na Ri lamh le Cruachain'. This clearly fixes the position of Rath Cruachain and Relig na Ri around the hill known as Sidhe Boidbh or Cruachan Boidbh, in the N of Ochaill Oichne. This is exactly where Boyhill (Cruach Boidbh) is today. There are local traditions about this area as described below and also evidence of the remains of several cairn tumuli (See Chapter 4). Further to this, legendary history places the burial mound of Queen Medb on the N part of Relig na Ri. This is corroborated on the ground as at the N of Boyhill is a burial mound surrounded by a ring of large stones known locally as Fert Medb.

Adjoining Mountain South is Slieveroe (Sli Bhrugh). The spelling of the name, as on the O.S. map, shows it was misunderstood as Slieve Rua or 'Red Mountain'. The total absence of any mountain or hill should have alerted the experts to the more startling significance of its true archaic name, which corroborates all the texts in ancient records referring to the character of this area surrounding Athenry. Long departed locals like Martin Kennedy, insisted emphatically that the O.S. toponymists recorded the incorrect form of the name, and that the original name was Sli Bhrugh, definitely not Slieve Rua. Traces of ancient burial cairns, denuded of their cairn fill, show the real significance of the original name, Sli Bhrugh (= the Necropolis Avenue), and corroborate the fact that this was part of the extensive Iron Age

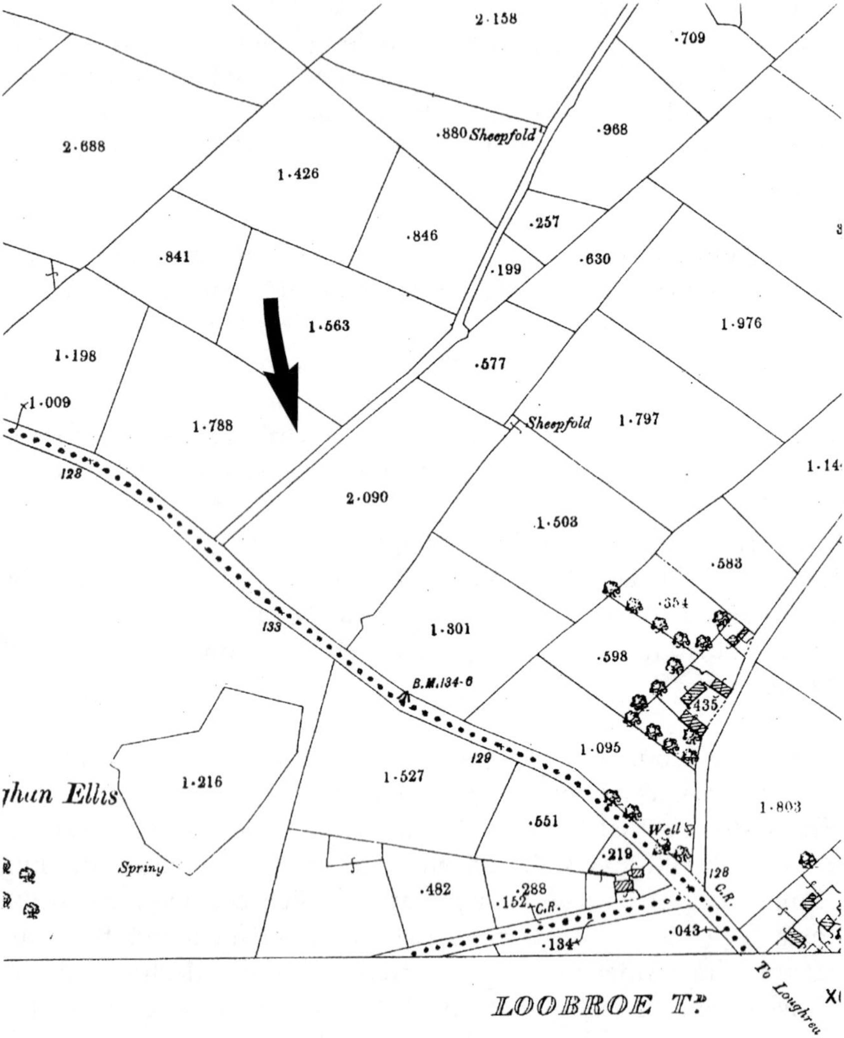

Figure 7.13. Bothar Bhodbh (Vov) starts from Boyhill and runs east and then north to Athenry.

Necropolis of Sidh Cruachain or Sidh Connaught. Those locals passed on a tradition that thousands of tons of stone were stripped from numerous cairn tumuli, including that of Queen Medb and of the Kings of ancient Ol nEgmacht (the original ancient name for Connaught). They were carted away by poor peasants under the heel of the Norman Lords of Medieval Athenry to build townwalls, streets, castles, churches and town houses. Scattered stones were collected from the destroyed tumuli in the now green fields of Athenry and piled onto 7 ft high by 10 ft wide stonewalls in the area which were formerly part of the outer ward defensive system. In many cases only the circles of kerbstones remain at ground surface to show where satellite cairns formerly stood. Only local farmers who have often broken ploughshares and the teeth and blades of hay machines on the barely protruding stones know these. As a result, farmers cleared the remains of many satellite mounds in an effort to remove these obstacles. Only a few on the W side of the Boy Hill have some cairn fill left to indicate what they originally were.

Local farmers claim a number of forts close to Athenry on the S and NW sides all went by the name of Rath Cruachu or Rath Cruachna. These included the one within the present parish cemetery, a double-ring stone caher a mile S of this in Clamper Park and another in Farranablake. There was also one known as Caher Cruachan Chonail Ceadgin, reputedly royal residence of Conal Cruachna, King of Connaught. This is the place where Conn Cedchathach, subsequent King of Connaught from whom Connaught was named (originally Ol nEcmacht), grew up together with his uncle Conal Chruachna. In the adjacent townland O.S. maps show a well, Tobar Chonaile. Only the Farranablake fort is given a garbled version of the name Cruacha on the O.S. maps.

Some 35 years ago several local elders such as Mr. Galvin from Slieveroe and the elder members of the Seises of Loobroo, claimed the whole area within a 2 mile radius of Athenry was known to "the old people long ago" as Rath Cruachu. Mr. Galvin noted that the Bottom road running EW along the S slope of Boy Hill was known as Bealach na Fert (Burial Mound avenue), mentioned in the tale of the Chase of the Magic Boar by Queen Medb. He pointed out sites of several stripped-down cairns, and the site of one called Carn Fraech, along side Bealach na Fert. He said that the Craughwell road running NS along the W side of Sli Bhrugh, Mountain South, and Boy Hill, to Athenry, was built on an ancient road known as Sli Bhrugh (Necropolis Road) which gave its name to the townland of Sli Bhrugh. He indicated the site of Clochar na ngFhotad (Palace of the kings of the Votadini vassals brought in by Lughaid Mac Con from NE Britain) in Moyode. He got his information from Martin Kennedy of Moyode who passed on the tradition that nearby Ailech Gorgin (now Rathgorgin) was named after Gorgin (Corc). He also noted that a massive stone fort across the river from Lios na Gal (foreigners' fort) at the E side of Moyode adjoining Killconierin was known as Cloghervaconierin (Clocher Mhac Chonierin), the Palace of Lughaid Mac Conn. Much of this is corroborated by archaeological remains and information in 'Dindsenchas Maigh Muc Dhruime', which tells that this plain was called Magh Fraech before it was renamed Magh Muc Dhruim. 'Aided Nathi' (an ancient legend) tells that 50 burial cairns dotted the plain of 'Relig na Ri lamh le Cruachain'. There are still some remains of cairn burial mounds in Boyhill and many others have been destroyed in recent years.

A piece of legendary history, taken from an early poem, suggests that Niall of the Nine Hostages had a connection with the Turoe area:

As we set forth from Cnoc na Dal
With the son of Eochail Muigh Mhain
More golden than the golden sobairche
Was the hair of the son of Cairinn

Eochail Muigh Mhain was father of Niall, Cairinn was his mother and they set forth from Cnoc na Dal (Knocknadala).

In a little-known Old Irish Law Tract 'Bech-Bretha' the word 'ecmacht' is used to refer to an 'enclave' or 'rampart-enclosed territory'. Significantly the older name for Connaught has been preserved in Ireland's oldest stratum of legendary history, the Ulidian Tales. This unique name, Coiced nOl nEcmacht, meas 'the fortress-enclave kingdom', or 'rampart-enclosed province'. In its earliest days the Turoe/Knocknadala/Rath Cruachu/Athenry oppidum was but a tiny embryonic enclave in the SW of the great Cruthin kingdom of the Northern Half of Ireland. It was this nucleus or enclave of the Belgae which gave ancient Connaught it's name, Coiced nOl nEcmacht. This name could be amended to 'Coiced nDhal nEcmacht' as in 'i nDhal Cuinn' or 'i nDhal Eogain', the names of Connaught and Munster respectively. 'nDhal' would sound like 'nOl' to a transcriber of the oral tradition who did not understand the Old Irish word 'nEgmacht'.

The Ulidian Tales, Ireland's oldest record generally accepted as an authentic picture of the archaic past, depict an Ireland dominated by two great powers constantly at war, Ulster and Connaught. These Tales are corroborated by Ptolemy's record in which his REGIA are located in Ulster and Connaught. This in turn is corroborated by the archaeological vestiges of Emain Macha in Ulster and further corroborated by the discovery of a vast defensive system centred around Turoe in Galway. This projects a political geography of archaic Ireland altogether at odds with that projected by current thinking on Irish history, which is based on Tara as the seat of the High Kings. It calls for a total review of Iron Age Ireland, dispensing with preconceived ideas and investigating properly the evidence becoming available in Galway.

Royal connotations in townland names and placenames

Townlands are among the oldest land divisions in Ireland. Often the names of the townlands reflect some historical or geographical feature of the area. Unfortunately, the origin of many townland names in Ireland has been lost. However, with a fresh look at historical events around the area, the townland names in Galway take on a new meaning. As we have seen throughout the text, these townland names relate to Royalty, earthen embankments, Celtic Tribes, etc. This is also true for the many place names and names for ringforts in the area.

Townland names:

Name	**Other name**	**Meaning**
Turoe	Cnoc Temhro	The same as the Irish name for Tara
Knocknadala	Cnoc na Dala	The only townland in Ireland with this name meaning 'Hill of Parliament'.
Fearta	Farty	Old Irish word for 'Graves', Royal Cemetery
Aberanville	Abar na bhFille	Field of the poets (filli)
Raford	Rath na bhFille Droichead	Bridge (droichead) of Poets Palace
Rathville		Rath (fort) of Mhil (son of Dela)
Cloghereavaun	Clogher Ui Mhaine	Palace of Maine, son of Queen Mehb
Attymonmore	Ait Ui Mhaine Mor	Place of the descendants of Maine Mor from Ulster
Clonkeen	Clan Cian	14 townlands of the Ciannachta
Clooncah	Cloon Cath	battlefield where the Ui Maine defeated Fir Belg
Ballinaulty	Baile na nUltaigh	area where Ui Maine Ulstermen settled
Dooughcloon	Doo cloon	earthen rampart field
Cahertinny	Caher Tinny	Caher of Tinny, early consort of Queen Meabh
Myode	Magh Fot(ad)	Plain of the Fotad (Votadini)

Kilascaul		Wood (coill) of A Scaul, a Celtic God
Brusk	Bresk	Brugh (Otherworld abode of) A Scaul
Boyhill	Cnoc Bodbh	Hill Og Bodbh
Loovroo	Lugh Bhrugh	Otherworld abode of the god Lug
Tobarnavean		Well of the Fianna
Sliverua	Sli Bhru	Otherworld cemetery

Ringfort names:

Name	Meaning
Cahertiny	Fort of Tinny, legendary King of Connaught
Caher Con	Fort of Con, King of Connaught
Caher Cormac	Fort of Cormac, King and grandson of Con
Clogher Ui Mhaine	Maine was son of Queen Medb
Caheroki	Fort of Eochaidh, father of Niall of the 9 Hostages
Clogher Nathi	Fort of Nath I, legendary King of Connaught
Caher Choinlle	King of Connaught, uncle of Conn of the 100 Battles
Domhnashatchell	Church of Satchell, a Bishop in Connaught.
Rathshatchell	Fort of Satchell
Clogher na gFotad	Fort of the Fotad (Votadini) Kings
Lis na Gal	Fort of the foreigners
Cahercinnmainmuigh	Fort at the head of the Plain of Main Magh
Rath Gorgin	Fort of Gorgin (Corc), the founder of Cashel

Local names:

Name	Townland
Cotni, Cetni	Part of Carrowkeel
Satchell	Part of Cloghereavaun
Taille	Part of Galboley, Fort of Tailliu
Cruachain Art	In Fearta, on the N slopes of Turoe
Loovroo	In Turoe and also near Athenry

Cruachain Sile	In Turoe and another near Athenry
Main Magh	Plain around Loughrea, including Turoe/Knocknadala area
Magh Muc Dhrium	Plain around Athenry (Muc Druim refers to the ramparts)
Cruachu/Cruachna	Area around Athenry
Suidh Boibh/CnocBoidbh	Boyhill, just outside Athenry
Ochaill Aichne	The forest area (ancient yew forest) between Clarinbridge and Athenry

These are but a few examples of the many townland names, fort names and other local placenames that relate to stories in folklore and events of early Irish history. While not being direct evidence, these do relate to historical events that occurred during the 1st/3rd Century A.D. and have enormous Celtic significance.

Celtic tribes reflected in townland names

In the Turoe/Athenry area there are many examples of townland names that reflect the presence of Celtic tribes. The following table lists a few of these.

Townland	Interpretation	Celtic tribe
Ballymannagh	Baile na Manach	Menappi
Newmannin	Mannin Uagh	Menappi
Mannin		Menappi
Manninard	Mannin Ard	Menappi
Ard na Manach		Menappi
Ganty	na gCantii	Cantii
Kilarrive	Coill Aithriabh	Atrebates
Carrownamorrissey	Caher na Marassaii	Marrasaii
Lisnamorrissey	Lis na Marassaii	Marassaii
Myode	Magh Fhoad	Fotads or Votadini
Treankyle	Trean Coill	Trenovantes
Ardnadomnan	Ard na Domnan	Dumnan

Many more such examples of Celtic tribes reflected in the townland names exist in the area. When added to the other evidence of Royal connotations, Celtic gods, embankment references etc. in the townland names, these becomes extremely significant.

Celtic gods and their influence on our current placenames

The pre-Christian Celts had many different gods. They had a very close relationship with their gods, who lived among them at an Otherworld Abode. Some of these gods were Cian (A Scail), Lug, Sile, Ethlenn and Boidbh. Significantly, these gods have left their mark on the placenames of East Galway. The following are just some examples of how these Celtic gods have left their mark on our townland names today. Sidh nEthlenn was the original name for Saint Ellen, a townland near Athenry. There is a hill in Turoe called Loovroo (Lugh Bhru; Otherworld abode of the god Lug). Significantly, the next townland to Boyhill (Relig na Ri) is also called Loobroo. Another Celtic god, A Scail, has left significant imprint near Esker. One townland there is called Killascaul (Coill A Scail), a lake there is called Loch A Scail and there is also Bru A Scail (Brusk). Close to these areas there is also Cruchain Aine and Cruchain Cian, called after two more Celtic gods. Cruchain Sile is also at Turoe and Suidh Boidbh is the name Boyhill is derived from. Indeed, there is an old road from Boyhill to the Holy Well and on into Athenry that is known to this day as Bothair Boidbh. Aggie Qualter in some of her writings referred to this road and described it as having lovely rounded banks. Unfortunately, it is now overgrown and stone walls have replaced the banks, but it can still, at least in part, be traced.

Similarity to other Celtic sites

One of the most imposing systems of Celtic fortifications in Central Europe, which was occupied by the Volcae or early Belgae, is the stronghold of *'Hradiste nad Savisti'* near Zbraslav in the Czech Republic. With *'Sance'* across the narrow valley from it, it occupies 170 hectares. The entire defended area within the outer ward is ringed in by a rampart and ditch, 6 metres high and 25 metres wide at the base, and some 9 km in length. The two sites, Hradiste and Sance, separated by the narrow valley, form one fortified entity, with extensive defensive earthworks running along the brows of the heights and thrown across the deep valley. Some 100 meters further down the slopes, is, what Jan Filip calls, the outer ward, *'Adamkovo Myto'*. Although so far removed from it physically, there is surprising similarity between this and the Turoe/Knocknadala oppidum. The two hills of Turoe and Knocknadala were similarly encompassed in one single entity by earthen ramparts thrown across the valley which separates them, and running along the brows of both hills. So too, approximately 100 yards further down the slopes of both hills another line of rampart ran parallel to that higher up. A further set of parallel ramparts stood at the lowest level. This set of staggered ramparts, linked by transverse ramparts, formed the inner ward around Turoe's sacred sites and the joint inner ward around the Turoe/Knocknadala complex. The densely populated area of the Hradiste sector descended in steps on the W side to the elevation *'U Altanu'* above the Vltava river. At the summit is *'Akropol'*, aptly describing the acropolis urban area. Similarly, Knocknadala hill descends to the bend on the Dunkellin (An Dughiortach/Dur) River with its sprawling urban complex, now destroyed, except for the site at Cotni. Just as Hradiste had its necropolis, so Turoe had its Fearta necropolis on its summit and N slopes.

The Celtic Oenach or Feis was always celebrated on an eminent broad hilltop, in the vicinity of the place of inauguration of Kings and the royal cemetery. The tribal rallying-places of the Celts were situated on heights, which were ancient burial grounds. The cults of the Celts were ancestor-worship and their feasts were commemorations, the sanctuaries were also fair-grounds. The population met at the political and religious centre of the tribes, which was the place where the tombs of its ancestors stood. It did this on feast-days, Samhain on 1st. November, Oimelc or Imbolc, on 1st. February, Beltaine, on 1st May, at the beginning of summer and Lugnassadh, on 1st August. These feasts were fairs, political or judicial assemblies, and an occasion for amusement and games, some of which, such as the races, were of religious origin (the horse races at Tailtiu). Turoe/Knocknadala has all these features of hilltops, burial grounds and ceremonial sites and can thus be compared to any other Celtic site. 'Knocknadala' literally means Hill of the Great Assemblies - The Old Irish word 'Dail', meaning 'great meeting place'. Fearta means 'Royal Cemetery', Aber na bFhille was a place for poets and what better ceremonial site than Turoe Hill with the Turoe Stone.

Ancient tales

Over the years the ancient tales of Ireland have been embellished with mythology. So much so that it can be difficult at this stage to tell fact from fiction. Adding this mythology to the tales was a clever mechanism used by Medieval writers to help transfer the traditions of Turoe to Tara. When put in the context of Tara the tales didn't quite fit and so were disguised to make them fit. When these tales are transferred back to a Turoe context they make better sense. The geography of a West of Ireland setting fits much better than the geography of an East of Ireland setting.

One of the better known Irish Tales is the Tain Bo Cuailne (the Wild Bull oof Cooley). In this tale, as understood at present, Queen Medb of Connaught, living at Rath Croghan in Roscommon, invaded Emain Mhacha because of her desire to get the Brown Bull of Cooley. The route she took from Rath Croghan to Emhain Maca was SE to Athlone and then NE to Armagh, a circuitous route. To explain this circuitous route, Fergus is supposed to have led Queen Medb astray in delaying tactics.

Now, let's suppose, as the evidence of this book suggests, that Queen Medb lived at Rath Cruacha of Athenry, then going to Athlone to get to Armagh makes perfect sense. In addition, two places that Queen Medb passed on the route to Athlone are in a direct line between Athenry and Athlone. Now there is no need for the embellishment of Fergus's behaviour.

Art, son of Conn of The Hundred Battles, was out hunting when he was informed of a massive invading army at Ath Cliadh Magh Ri led by Lughaid MacCon. The next day Battle between Art and Lughaid ensued at Mamucrimme (Magh Muc Dhruime), Art was slain and Lughaid Mac Con assumed the Overkingship of the Fir Belg. Placing Art as Overking at Tara, as is the general account of this battle, raises some very basic geographic questions. If Art was at Tara, why did Lughaid Mac Con invade from the West and endanger his army on a long sea journey? What was Art doing hunting near Clarinbridge and how did he get his army there before the invaders were hardly landed? Now, let's suppose, as the evidence in this book suggests, Art reigned at Turoe, then it is perfectly plausible that the invading forces would use Ath Cliadh Magh Ri and also possible that Art and his army would be there quickly.

Further to this, dying tradition at Clarinbridge and at Turoe, 15 miles inland, asserts that Art was buried where he was slain, near Clarinbridge. When Cormac, son of Art, was restored to the throne he had his father's remains removed and buried at Turoe. Cruchain Art, the burial mound of Art, was testimony to this until it was unceremoniously removed recently. In pseudo-historic documents Art is supposed to be buried at 'Treoit', near Tara of Meath. However, substituting this for 'Turoit' (meaning Turoe) was very easy. Thus with a simple stroke of a pen, the burial place of Art was moved to the Tara district.

The story of the death of Queen Medb also fits well into a Galway setting. Medb had her sister Clothra assasinated by the Mairtine of Munster and took over the throne of Connaught. Clothra is said in pseudo-history to have had a Royal Crannogue on Lough Ree (near Athlone), but legendary history states that Medb took over her sisters crannogue on Loughrea Lake, where several crannogues exist. An 11th Century text claims that Furbaidhe was saved from Clothera's womb by caesarian section and that he awaited his opportunity to gain revenge for his mother. Before he died, King Conor Mac Neassa of Ulster had put Furbaidhe under 'geish' to assassinate Medb whenever he got the chance. One day while Medb bathed at the lake he assassinated her. Main, Medb's son took over the throne and had his Palace at Clogher Ui Mhaine (near Kiltullagh). The plain North of Loughrea was, until reletively recent times, called Main Magh (the Plain of Main). Main had a sister called Fionnura, who later reigned as Queen of Connaught. Finnure Hill, 5 miles ESE of Turoe, is called after her.

So facts surrounding the life and death of Queen Medb fit easily into the setting of Turoe and Athenry.

In the area around Athenry/Turoe there are many folklore tales about the Fianna fighting men. In addition to this, there are many townland names that have come from the Fianna. These include Tobernavean. The Fianna were supposed to be the army defending the Overkings that lived at Tara of Meath. However, the extent of the folklore about their presence, particularly in the Loughrea, Turoe, Knocknadala, Slieve Achta Mountain area, raises some very basic questions. If the Overkings lived at Turoe, then the presence of the Fianna in the Achta Mountains makes perfect sense.

There are many other legendary battles and tales that have been implicated as having a Tara setting. These include Cath Boindi Andso, Cath Airtech, Bruidhen Da Choca, Togail Da Dearga, Cath Ruis Na Rig etc, etc. They don't quite fit the Tara setting that has been given to them and a close scrutiny of the texts raises some serious questions that have been, for the most part, glossed over. However, when these are placed in a West of Ireland setting, centered on Turoe/Athenry, they appear in a whole new light and begin to make perfect sense. Detail of these is not given here, but is discussed in detail in 'Hand of History, Burden of Pseudohistory.

Chapter 8

The fall of Turoe/Knocknadala and Athenry as Centres of Power and the suppression of their existence

Turoe/Knocknadala and Athenry were eventually to fall as Centres of Power in Connaught. As if this was not bad enough, the very existence of these important areas was to be wiped from the history books. An understanding of how and why this was done will come from a brief look at the events surrounding the times when history began to be recorded in writing about the 7th Century.

The end of Turoe/Knocknadala and Athenry as a power centres

After several hundred years of expansion from the core of Turoe/Knocknadala, the Fir Belg tribes extended to the furthest reaches of Connaught. During the reign of Eochaidh Muigh Mhain (that is from Main Magh, the Plain around Loughrea, centred on the core of the Turoe/Knocknadala oppidum) in the 5th Century this expansion was continuing. Eochaidh had three legitimate sons Briun, Fiachra and Ailill. His fourth and youngest son, Niall, was born of a British slave girl, Cairinn Cruithnech. Her name indicates either that she was of the Cruthin people of Britain, that is, she was a Scottish Pict or simply from Scotland. The fact that Niall was the son of a slave girl meant that he was illegitimate and ineligible to the throne. The younger sons of kings became the commanders of the fighting forces of the king. Legendary history tells that Niall was sent N against the Ulstermen. As leader of Connaught's fighting men, Niall cut loose from Briun's overlordship (Briun had become Overking on the death of Eochaidh) and gained land for himself from the lands of the Cruthin of NW Ulster. He later became known as Niall of the 9 Hostages because, by force or otherwise, he persuaded the 9 Airghialla tribes (the Eastern Subjects of Briun) to turn against Briun and become subject to himself instead. In the same way he persuaded his brothers Fiachra and Ailill to withdraw their followers from Briun's overkingship and ally themselves to Niall instead.

At this point it is important to remember that Eochaidh Muigh Mhain, who certainly lived around Main Magh, may have lived at Cloghereavaun. There was a stone fort there called Caheroki, which may have been the place where he lived. This is corroborated in a legendary poems which tell of his residence near Knocknadala:

When we set forth from Cnoc na Dal

with the son of Eochail Muigh Mhain

More golden than the golden sobairche

was the hair of the son of Cairinn

A second poem deals with the rough treatment of Cairinn and her son Niall by the jealous wife of Eochu Muigh Mhain. This glorifies Niall as 'the champion of Maen Magh' (the plain around Turoe in Galway).

After the death of Eochaid Muigh Mhain, Briun succeeded him as Overking of the Belgic tribes while Fiachra became the local King of Connaught. Fiachra marched northwards side by side with Niall and extended his kingdom of Ui Fiachrach up through North Galway and Mayo. Like Niall, Fiachra now refused to submit his expanding kingdom to Briun's Overkingship. Briun was away on an expedition against the Cruthintuatha Croghan of NE Connaught, leaving his Turoe/Knocknadala Power Centre poorly defended. Realising this, Maine Mor, son of Niall, led a massive army of Ulstermen and the Airghialla (who had submitted to Niall), south against Briun's overkingdom on both sides of the Shannon. Maine Mor's invasion is recorded in the 'Life of Saint Grellan'. The route of Maine Mor's army west of the Shannon led through the 6 Sogains, isolated Cruthin lands in the area of Tiaquin. Cian, King of the local Clann Cian townlands was left behind by Briun to defend the area and keep a check on these Sogain townlands. Maine Mor's army passed through Bealach na nArm to Knockaboy near Attymon, almost to the edge of the inner ward of the Turoe/Knocknadala oppidum.

The area where Maine Mor met resistance and engaged in battle is forever etched on the townland names of that area. This area is none other than Attymon. The townland name is Attymonmore (Ait Ui Maine More). This is the place where the followers of Maine Mor settled after their victory. Similar settlement took place a little further West in what is now Balinaulty (Baile na nUlaigh; town of the Ulstermen). The battle between Maine Mor and the last line of defense of the Fir Belg took place at Clooncath (Cloon Cath; Field of battle). Indeed many locals hold the Gortnalone is pronounced as Gort na Lan (Field of Spears), again recording a battle. Legend holds that the defending Fir Belg were annihilated in this battle and that 'the ground opened up to swallow them', suggesting that many of them were lost in the bog. Numerous finds in the area corroborate this. Farmers cutting turf have frequently encountered masses of bones, perhaps a burial site. Others have encountered wooden stakes and spears in the bog. This was also the location where the Attymon Hoard (a decorative La Tene horsebit) was found. Who knows what other treasures lie beneath the bog here.

After Maine Mor's invasion narrowed the territories of Briun and encroached dangerously on his Turoe/Knocknadala power centre, Briun cut a wide wedge north through the heart of Fiachra's kingdom dividing it into north and south sections. This cut off the link Fiachra had with Niall and his sons.* Briun abandoned the Turoe/Knocknadala area and set up the new centre of power on a crannogue on Loch Hacket 6 miles SW of Tuam in the territory of Magh Seolgha (Figure 8.1 - 8.3). From here further disputes between Briun and Fiachra were fought out. Briun and his descendants expanded from there and began to push the centre of Fiachra's kingdom further and further south from Rath Cruacha of Athenry. That is why Nath I (Dathi) the son of Fiachra and next King of Connaught, had his royal centre at Toor Cloghar Nath I beside Blackgarden and Seefin south of Craughwell. The later Ui Fiachrach King, Guaire, had to move his palace south to Dun Guaire in Kinvarra (Figure 8.4). That is how it came about that there were 2 widely separated Ui Fiachra kingdoms in Connaught, one in the SW of Galway and the other up in W Mayo and into W Sligo.

In 784 a Northern Ui Briuin king, having established his dynasty at Rath Croghan of Roscommon, devastated his SW Galway Ui Fiachra rivals for Connaught's Kingship. To add insult to injury he raided and ransacked Medb's palace and tumulus at Rath

Cruacha of Athenry which the Ui Fiachra upheld as the ancient Capital of Connaught, thus erazing it from popular tribal memory in favour of Rath Croghan of Roscommon. To crown this outrage he desecrated Medb's royal crannogue on Loughrea Lake and concocted the claim that her crannogue had been on Lough Ree and not Lough Rea (a simple change of one letter).

Niall of the Nine Hostages remained in Ulster where he went on to be the Father of the great Ui Neill dynasty, which took over Tara in the 7th Century and influenced much of early Medieval history. Until that time Tara remained in the hands of the Ulster Cruthin. Several texts, including the Book of Leinster, corroborate this fact.

Supression of the true history of Turoe/Knocknadala and Athenry

If such an extensive Centre of Power existed at Turoe/Athenry, the big question is, why is there nothing in the history books about it?

Figure 8.1. After moving from the Turoe/Knocknadala area, Briun set up his Royal crannogue at Lough Seola (Lough Hacket today).

Figure 8.2. Cairn Cesra on the summit of Cruach Magh Seola, overlooking Lough Seola (Lough Hacket), 6 miles SW of Tuam. Briun set up his new Centre of Power here when he abandoned Turoe.

Figure 8.3. Linear embankments on the north slopes of Cruach Magh Seola.

Figure 8.4. Dun Guaire Castle in Kinvarra. In the foreground are the ruins of the Royal crannogue of Guaire, 6th century King of Connaught. Guaire was an Ui Fiachra King who vied with the Ui Briun dynasty for Connaught's Overkingship.

The simple answer appears to be that the existence of these Centres was suppressed in order to glorify the Centres at Tara of Meath and Rath Croghan of Roscommon. The Ui Neill dynasty had become dominant and in order to gain more power and authority they wanted to give the impression that Tara had been the seat of their ancestors from time immemorial. Thus, they were involved in a clever campaign of forgery and invention, which has been carried on to this day. Although this sounds unbelievable, it is widely accepted that many of the earlier historic documents have been invented or altered. Since the Abbots and scribes of the monasteries were the younger sons of the great kings, they concocted the great Annals of History to provide political propaganda glorifying their own royal dynasties, and their elder-brother Kings who supported them financially. They invented many of the great books that are understood to be the basis of our history and altered many others. Thus, the traditions of Turoe were transferred to Tara and those of Athenry to Rath Croghan, creating the myth that Tara was the seat of the High Kings of Ireland and Rath Croghan the Capital of Connaught. At the same time they wiped Turoe and Athenry almost completely off the Iron Age map of Ireland.

At the time there was a power struggle within the Church in Ireland. On the one hand there were the independent Churches in each 'tutha' or area. These operated by their own guidelines, but a great effort was made to get these independent Churches to support St. Columbcille and his monastic system. On the other hand there was the ecclesiastical system where the Church at Armagh tried to organise the Church under its rules. Because the independent Churches were associated with family dynasties, there was an ulterior motive in winning a Church for Armagh. Along with it came the support of the ruling family and thus the support of the inhabitants of an area and their financial contributions. So there was quite a lot at stake in this struggle for supremacy. The Church at Armagh, which supported the Ui Neill Dynasty, eventually won supremacy. They were thus in a position to spread their invented history to the rest of Ireland.

The evidence for suppression of the Turoe/Athenry Centres is very complex, but it is to be found in many ancient documents. In general, these documents have been interpreted with the preconceived notion that Tara was the seat of the High Kings during the Iron Age. In this interpretation there are many inaccuracies, which have been overlooked, and historians over the years have adapted much of Tara's history to suit the documents. However, if these ancient documents are interpreted without the preconceived notion of Tara and with the knowledge presented in this book, it becomes immediately apparent that they perfectly suit a West of Ireland setting. Many of the inaccuracies are thus explained and many questions answered. Without any intervention there is strong evidence in the Turoe/Athenry area for all the Kings and places that are associated with Tara. This has been demonstrated in the placenames, oral history and archaeology, as detailed in this book.

Although not dealt with here, the evidence that ancient documents have been tampered with and misinterpreted is fully reviewed and explained in the forthcoming book 'Hand of History-Burden of Pseudohistory'.

Current evidence is beginning to suggest that, contrary to the popularly held idea, Tara was not the seat of the High Kings of Ireland during the Iron Age. This begs the question: Where did the Kings who were supposed to have lived at Tara actually live? When the preconceived idea that 'Tara was the Capital of Ireland during the Iron Age' is dropped and the current evidence taken into account, then a new interpretation of Irish Iron Age history becomes necessary. Hopefully, this book will inspire this reinterpretation, and that the evidence presented in it will be taken into account to build a real picture of Ireland during the Iron Age.

The work detailed in this book is ongoing research. If anybody has any comment or additional relevant information I would be delighted to hear from them. I can be contacted at Strawhall, Fermoy, Co. Cork or by email at:
jordankc@eircom.net
Comments can also be made through the web site at :
www.turoeandathenry.com

Further reading:

Michael Herity and George Eogan, Ireland in prehistory, 1989, Routledge.

Michael Ryan (Ed), The Illustrated Archaeology of Ireland, 1991, Town and Country House.

James Charles Roy, 'The Road Wet, The Wind Close - Celtic Ireland', 1987, Gill and McMillan.

F. T. O Rahilly, 'Early Irish History and Mythology', 1984, Dublin Institute for Advanced Studies.

Francis John Byrne, 'Irish Kings and High Kings', 2001, Four Courts Press.

Nora Chadwick, 'The Celts', 1976, Penguin.

Ciaran Brady (Ed) 'Interpreting Irish History', 1994, Irish Academic Press.

John Waddell, 'The Prehistoric Archaeology of Ireland', 2000, Wordwell.

Barry Raftery, 'Pagan Celtic Ireland: The Enigma of the Irish Iron Age', 1994, Thames and Hudson.

Jan Filip, 'Celtic Civilization and its Heritage', 1977, Collet.

P. Berresford Ellis, 'The Celtic Empire', 1992, Constable.

Barry Cunliffe, 'The Regni', 1973, Duckworth.

Norman Mongan, 'The Menapia Quest', 1995, Herodotus Press.

D.W. Harding, 'The Iron Age in Lowland Britain', 1985, Routledge.

S. P. O Riordain, 'Antiquities of the Irish Countryside', 1953, Metheun.

Barry Raftery (Ed), 'Atlas of the Celts', 2001 London

Daibhi O Croinin's 'Early Medieval Ireland, 400 - 1200', 1995, Longman.

R. F. Foster 'The Oxford Illustrated History of Ireland', 1989, Oxford University Press.

T. W Moody and F. X. Martin 'The Course of Irish History', 1967, Mercier Press.

Barry Cunliffe 'The Ancient Celts', 1997, Oxford University Press.

Barry Cunliffe 'Oppida in Barbarian Europe', 1976, Oxford University Press.

Donnchadh O'Corrain 'Ireland before the Normans', 2000, Four Courts Press.

Myles Dillon 'Early Irish Literature', 1948, Chicago.